Handbook of Energy Agriculture

Handbook of Energy Agriculture

Hazem Shawky Fouda

Delve
PUBLISHING

www.delvepublishing.com

Handbook of Energy Agriculture

Hazem Shawky Fouda

Delve Publishing
224 Shoreacres Road
Burlington, ON L7L 2H2
Canada
www.delvepublishing.com
Email: orders@arclereducation.com

ISBN: 978-1-77469-081-9 (Hardcover)

ABOUT THE AUTHOR

Hazem Shawky Fouda has a PhD. In Agriculture Sciences, obtained his PhD. from the Faculty of Agriculture, Alexandria University, 2008, MSc. In Agriculture Sciences from the Faculty of Agriculture, Alexandria University in 2004, Post-Grade Diploma in Cotton, 2001, BSc. in Agriculture Sciences, from the Faculty of Agriculture, Alexandria University, 1997, He worked in Cotton Arbitration & Testing General Organization (CATGO) since 1999 till 2018. Was working in the International Cotton Training Center (ICTC) – Cotton Arbitration & Testing General Organization (CATGO) from 2000 till 2015, as a Lecturer & Classer's Trainer for Egyptian and foreigner classers, technicians, ginners, spinners & traders in all cotton aspects starting from picking seed cotton till preparing bales for export including other cotton aspects and training them practically on classing cotton. Besides that he was an editor and active member in the Research & Translation Committee, participating in issuing weekly, monthly and annually issues about the international & local cotton market including price trends and direction, recent developments & researches concerning cotton production, protection, harvesting, ginning, fiber testing, spinning & weaving since its foundation in 2000 till 2014 and since 2015 till 2018 he worked as an inspector, from 2018 till present works as a consultant.

TABLE OF CONTENTS

LIST OF FIGURES

ACKNOWLEDGMENT

First and foremost, I feel always indebted to ALLAH, the most beneficent and merciful.

Words stand short when they came to express my gratefulness to my late parents Shawky Fouda and Fatma El-Hussainy, my uncle Ahmed Fouda, my wife Heba Sadek, my son Mohammed Fouda, my daughter Raghad Fouda,

And a special thanks to my sister Riham Fouda, relatives and friends for their support, encouragement, care and patience and for their love.

PREFACE

The progress of the global economy and innovation can be linked to the development of new fuel sources. In the 19th century, for example, before the modern industrial revolution, coal was the main source of fuel but it was later replaced with oil and combustible gases. Before the end of the 20th century, many fuel sources were discovered, such as sunlight, wind energy, hydroelectric power, and biofuels. Nowadays, with the improvement of professional technology, it is also conceivable to obtain energy from atomic splitting.

This volume discusses global energy use patterns and energy needs for production lines, businesses, vehicles and airplanes, and domestic use. It is estimated that annual energy use is growing at a rate of 2.3%. This is mainly due to human activities and the need to develop, which is reflected in the higher demand for industry, the increase in transportation systems and new occupations that all require energy to operate.

The topics covered in this volume include chapter 1: energy availability and energy trends, chapter 2: usage of energy in the production of agriculture, chapter 3: choosing the right energy source for agriculture, chapter 4: energy options in irrigation, Chapter 5: Energy Options In Harvesting Agricultural Production, Chapter 6: Energy For Drying Of Agricultural Production, Chapter 7: Cooling And Cold Storage Energy For Agricultural Production, chapter 8: challenges on agriculture energy use, chapter 9: policies and regulations for the energy-agriculture, and chapter 10: conservation of agricultural energy.

Energy has always been the driving force of development, which has promoted the demand for sustainable sources of power in the transportation, agricultural and manufacturing industries. The volume explores modern energy consumption trends and future prospects of the sector. The authors argue that to maintain sustainability and effective energy use, it is essential to adapt to renewable and cleaner energy sources.

The energy sector should cover life cycle analysis to help choose and implement a more sustainable energy framework and develop a better and smarter energy matrix, of which the energy capacity dynamics is a basic component. These inquiries will be discussed in the volume, focusing on the existing situation, problems and possible solutions, as well as the anticipated future of energy.

The volume mentions that renewable energy-solar, water, wind, biomass-are inadequate to meet modern energy needs. While the use of oil, coal, and natural gas provides sufficient energy for industry and transportation, they cause environmental pollution. In the endless cycle of development, the use of fossil fuels (FF) and fossil-derived energy (FDE) is subject to availability and natural reserves which are limited.

The volume recommends using different methods to solve the energy availability problem: 1) reduce the required energy needs by improving energy efficiency through rational distribution and utilization; 2) use alternative renewable sources of energy, such as nuclear power 3) Develop improved and efficient energy storage structures. In addition, energy decision makers must provide better policies to study and understand the lifecycle of energy creation and consumption for better management: from the extraction of crude oil raw materials, transportation, refinement and ultimate use of the product.

Chapter 1

Energy Availability and Energy Trends

CONTENTS

1.1 INTRODUCTION

Energy demand is expected to increase sharply in the coming years as a result of population growth and economic development. When their markets transition from livelihoods to manufacturing or services, many people around the world are undergoing significant lifestyle shifts. Developing nations will see a large rise in energy demand, with global energy use projections increasing from 46% to 58% between 2004 and 2030. However, per capita intake estimates are expected to be far below the average of the Organization for International Cooperation and Development (OECD) countries. Energy demand in developed countries is expected to rise by 3% annually from 2004 to 2020. In developed countries where domestic economies are mature and population growth is predicted to be relatively strong, energy consumption is estimated to rise by less than 0.9% per year, although from a much wider viewpoint (Armaroli & Balzani, 2007). Energy demand is projected to be higher by 2010 than that of industrialized regions in developing regions. By 2030, about half of the world's energy demand for electricity and one quarter for transport, primarily in the form of petroleum fuel, is expected to rise. Energy demand growth would be mainly due to rapid economic growth in Asian economies, especially China and India. Energy demand is expected to grow at an average annual rate of 3.7% in developed countries in Asia, far higher than any other region. Over the next 20 years, Asia will more than double its energy usage and is projected to account for nearly 65 percent of the overall energy demand increase of all developing countries. While energy demand in developing countries in regions other than Asia is projected to rise more slowly, it still exceeds the global average. While all regions will play their part in future energy supply and demand, the region will have a major interest in future energy output due to the strong consumption growth anticipated in Asia. Many of the world's non-renewable electricity comes from oil, coal and gas in particular. Just over 13 percent of the world's electricity comes from sustainable energy sources, 10.6 percent from conventional sources from existing energy sources and municipal waste. Hydro, geothermal, hydro, wind, tidal and wave sources make up the remaining renewable energy options. Full global energy demand forecasts indicate that the bulk of fossil fuels will rise in absolute terms between 2004 and 2030, with relatively small contributions from nuclear and other sources. With changes of between 65 and 74%, gas and coal are expected to change in percentage terms. The use of oil is expected to increase by 42%, while the use of nuclear and renewable energy is expected to increase by

44% and 61%, beginning from a much lower starting point (Anishin et al., 2013). The final contribution from several sources would, for the most part, be focused on policy recommendations. For this function, forecasting may also be used as a starting point for further debate.

1.2 RENEWABLE ENERGY

Renewable energy is generated and/or generated from sources that can be indefinitely renewed or produced in a renewable manner, such as biomass, hydro, solar and wind power. The use of renewable energy sources is expected to increase, given the anticipated dominance of fossil fuels. Company renewables will grow at an annual rate of about 1.9% over the coming decades, based on forecasts by the US Energy Information Administration (EIA). The largest absolute growth is projected in North America, the developed countries of Asia and Central and South America. In the Middle East, Asia's developing countries and Central and South America are expected to see the highest average rates for renewable energy use. As in Central and South America, the trend is motivated more by increasing electricity demand in advanced Asian countries than by reliance on renewable energy. It is expected that the share of electricity from funded renewable sources will rise in most regions of the world in the coming years. Total use of renewable energy is by far the highest in Central and South America, where commercial, non-fossil fuel alternatives are now well known (Ahmad & Zhang, 2020).

Africa and Latin America are the two most important renewable energy regions. The use of wood to heat and cook in Africa is mainly due to this. The strong use of renewables in Brazil accounts for 45 percent of all renewables in Latin America, including hydroelectricity, biomass and sugar cane ethanol. Biofuel consumption, with the notable exception of that of the Russian Federation, is on the increase in the majority of G8+5 countries that use the highest amount of energy in the world. The US, China and India use by far the highest quantity of biofuels in absolute terms. Bioenergy grew as a proportion of overall energy consumption between 2000 and 2005, giving all economic advantages to bioenergy in Germany, Italy, the United Kingdom, the United States and Brazil. However, the relative use of biofuels in China and India has been decreasing, with rising prices of fossil fuels surpassing strong economic growth rates.

Figure 1.1: The key trends that will shape renewable energy.

Source: https://www.weforum.org/agenda/2018/01/clean-energy-renewable-growth-sustainable-key-trends/

1.3 BIOFUELS FOR TRANSPORT IN BRAZIL

Liquid biofuels account for just about 1% of transport fuel consumption worldwide. Brazil is noteworthy with the exception of this average. During the first global oil crisis of 1975, Brazil initiated a national biofuel program that culminated in the generation of ethanol from domestic sugar supplies on a wide scale. In Brazil, more than 90% of all cars manufactured and sold are "flex," equipped with ethanol or petrol engines, as well as mixtures. Brazil has recently unveiled a global initiative to encourage biofuels as a safe solution to transporting fossil fuels.

Brazil's sugar cane biofuel price is cheaper than petrol, where it's more than USD 35 a barrel (Basore et al., 2018). By comparison, US corn bioethanol is competitive at a cost of USD 55 per gallon, and bioethanol needs to be competitive in the European Union at a price of USD 75 to USD 100 per barrel of crude oil. Biofuels in Brazil have been largely popular due to the high production of sugar cane and the adequacy of food for the successful conversion of ethanol. About 190,000 hectares of annual sugar cane plantations are planted, mainly from southern parts of the country. Brazil is expected to be the world's leading exporter of biofuels.

1.4 WOOD-BASED ENERGY

In the future, the global distribution of biomass and its potential to substitute oil as biofuel remain unclear. Global timber production in 2005 amounted to approximately 1.7 billion cubic feet or approximately 1.8 billion cubic meters. Almost 65 percent of the world's commercial round wood has been harvested in developing countries, compared to just 13 percent of the world's timber. India (306 million cubic meters), China (191 million cubic meters) and Brazil are the main suppliers of carbon wood (138 million cubic meters). Just a few developing countries are involved in the processing of fuel wood, including the USA, Mexico, Finland, Sweden and Austria. However, there are data entry problems, and studies of household fuel use have demonstrated substantial usage in many other developing countries. The vast majority of fuel wood is domestically grown and used. Given that fuel wood is mainly used in private homes and is widely traded informally, sound national statistics are difficult to obtain. Numerous additional warnings apply to the detailed and realistic effects of wood fuel. Bioenergy has historically been the most productive form of biomass. Since the fire started, wood has been used for heating and cooking purposes. Sometimes used in developing countries for agricultural purposes, such as drying fish, tobacco production and bakery bricks. It is primarily used for power production in the timber industry in developed countries (Asif & Muneer, 2007). In recent years, wood energy has become an ecological alternative to fossil energy, and investments have been made to increase efficiency, especially in packaging, heating and power applications. Climate change has allowed wood-based energy systems to be developed in many parts of the world. New technologies, especially in highly forested countries with developed wood processing industries, are improving the economic viability of wood-based energy production. In absolute terms, the United States, Canada, Sweden and Finland are the largest industrial users of bioenergy timber in the OECD. A significant proportion of wood biomass used for energy, including wood pulp and wood residues, is recovered from indirect sources in these regions. In both countries, industrial applications accounted for a little over 50% of the overall consumption of bioenergy. Fuel wood is the predominant form of wood energy in rural areas in most developing countries, while carbon dioxide is an important source of energy in many urban dwellings in Africa, Asia and Latin America. About 90% of the world's wood fuel consumption (wood and charcoal fuel) remains in developed countries, and wood is also the primary source of cooking and heating oil. Global wood fuel consumption has been largely stable over the last 15 years, ranging from 1.8 billion to 1.9 billion cubic meters. The increasing use of wood oil is mostly due to the rise in African use, as the world trend shows (Batterberry et al., 2010). By contrast, in Asia and Oceania, non-OECD countries are facing a downward trend as income grows and urbanization progresses. Recent EU proposals to increase the share of renewable energy in total energy usage to 20% by 2020 could result in higher future demand in OECD countries in Europe.

Owing to the difficulties of gathering correct data on the use of wood fuel, the analysis of the findings is wise. Recent increases in international oil costs, for example, have limited the speed at which wood fuel customers have moved to cleaner and more efficient heating and cooking fuels.

Figure 1.2: The Different Uses of Wood Biomass Energy.

Source: https://www.planete-energies.com/en/medias/close/different-uses-wood-energy

1.5 IMPEDIMENTS TO ACCURATE WOODFUEL INFORMATION

It has always been difficult to establish a comparative awareness of wood-based fuel consumption.

The most important things are as follows:

- High-intensity surveys are critical for accurate statistical results, as wood fuel production and usage vary greatly in different areas and periods of the year.
- Wood fuel is supplied exclusively to the collector's own needs and is not sold for data collection at any venue, such as supermarkets, shops or warehouses.
- The economic value of the sector in the majority of countries is low, and thus statistical expenditure is considered not too beneficial due to its low price in most countries.
- Many countries do not have the financial and human resources required to gather information on wood fuel, particularly if the poorest countries still need the most wood fuel.
- Sector-interesting entities (e.g. government agencies dealing with agriculture, forestry, energy and rural development) are often poorly structured and the importance of knowledge gathered by a single agency can be insufficient.
- Many government forestry departments focus their energies on commercial timber production and neglect non-commercial tree production.
- Basic definitions, measurement conventions and transition factors for wood fuel details are lacking, making it impossible to compare data between regions and over time.
- The extent of wood residues available for energy use may be underestimated as a result of widespread illicit deforestation (Badawy et al., 2010).

1.6 NEW TRENDS IN ENERGY PRODUCTION AND UTILIZATION

Electricity has long been associated with the development of nations since ancient times when people started to use fire for hot, cooking or defensive purposes. Energy connectivity has encouraged the demand and expansion of the population.

The energy needs of light, water, wind, biomass, the most conventional and renewable energy sources have not been met. The use of coal, gas and oil allows larger volumes to be used, for contemporary use, for industrial production or mobility/transport, for environment and ventilation, for communications, etc. However, this infinite cycle, in which people require money to advance and development requires the use of increased volumes of readily available oil, has its disadvantages: the use of fossil fuels (FF) and fossil-based energy (FDE) is restricted and the use of reservists is not similarly common worldwide; the use of FF and F Theory (FDE) has been limited. In comparison, the question of usable power has been discussed in three key approaches: (1) the amount of electrical resources needed to improve energy supply and usage efficiency (smart grids); (2) the use of alternative energy sources, in particular, nuclear energy (though major risks have been established in recent years, and are therefore not available anywhere) or, in particular, nuclear energy. This section, which demonstrates the recent trends in the energy market, also offers a short summary of all these approaches and their associated ties (Cardenas et al., 2016).

1.6.1. Energy Demand

Indeed, since the Economic Revolution after the Second World War and the boom in energy consumption, global demand has risen considerably. This has led to a steady rise in energy use worldwide over the past 50 years. The highest fraction is FDE: oil with a rise of more than 280 percent, led by coal and natural gas. Around the same time, hydroelectric capacity increased by 500 percent. However, the percentage of fossil fuels is much lower. Energy production has risen significantly, but has stabilized in the last five years, possibly after nuclear power plant accidents such as the one that happened in Fukushima in 2011, leading to increased public awareness of the risks involved. There has been an improvement in the use of such natural resources over the last 10 years. However, its share is still marginal (< 7%) relative to any other source of electricity. Population growth in developed countries such as China and India means that surplus energy consumption is much higher globally. Measures must then be taken to reverse the condition.

1.6.2. Energy Sources and Problems

The primary sources of electricity in the last 50 years have been coal, oil and gas, as previously stated. They are controllable but not widely distributed or available in infinite amounts. As far as nuclear power plants are concerned, while this is one of the most innovative and reliable energy technologies, little can really be done to stop a global catastrophe through an accident (Ceylan et al., 2004). The new tsunami in Fukushima has forced global policymakers to revise their energy plan. It reversed the decision to delay the closure of a number of nuclear power plants, and Japan decided to change its energy policy by relying almost exclusively on nuclear energy. Consensus has already been established on the long-term wasteful essence of current energy consumption and manufacturing activities. Today, and thus, the future sources of energy would be greener. In fact, the three major facets of sustainability, natural, economic and social, should be green and have a less negative effect. Renewable electricity supply is, in reality, one of the essential prerequisites for sustainable industrial growth. It is acknowledged at the governmental, institutional and even corporate level, and all players are committed to producing more productive RES and expanding their share of the global energy mix.

Strong investment is currently being made in either basic or applied science, or most countries and/or foreign bodies (i.e. the EU) have some kind of policies or initiatives to further improve and implement RES. Two examples are the energy transition policies of the European Union and the Energy 2020 plan and the 2050 vision.

Overall, the influence of the new RES on the energy balance remains marginal. While some regional success stories like bioethanol are available in Brazil, further work is still required. The essence and role of the current RES is the cause of all of this condition. In fact, the majority of RESs are restricted in architecture, target markets and operating conditions, and only limited environments can use induced electricity. On the other hand, they rely on conditions outside the control of the producers on the local environment and other external factors. Examples include: atmospheric constraints, environmental requirements that could limit acceptable solutions (e.g. hydro generation requires water availability), high building costs, and the use of such parameters may have substantial possible environmental, social and economic implications (biofuels vs. food disputes). Furthermore, biomass and biofuels are also disadvantaged because, while their supplies can be managed, their supply and development can also interfere with the use of arable land or food and feed (Cook, 2021). Furthermore, energy inefficiency remains important and must be a focal point for heavy expenditure. However, these issues continue to be resolved and the RES hopes to cover a wide variety of energy needs in the future. Today, it is recognized that a variety of RESs are needed to ensure that their intrinsic presence and characteristics have an effect on global energy supplies. As a result, current RESs are badly integrated and even worse integrated with other systems, especially industrial manufacturing.

1.6.3. Nations Commitment Towards Sustainability

The Brundt Land report was published in 1987 to address climate and construction as a single topic and was the key point when environmental concerns were first and foremost on the national agenda. In view of the fact that (a) global warming has taken place, (b) man-made CO2 emissions have been developed, the Kyoto Protocol has led to the reduction of GGE emissions by all 37 signatory developing countries and the European Community. Global policymakers have agreed that something has to be done to shift the way of thought and shifting transformation patterns, understand the human impacts on the atmosphere and, in particular, the tremendous rise in CO2 pollution and the degradation of the ozone layer caused by the use of FDE. A number of efforts that were not universally recognized and dedicated to their political commitment to fostering a stable future at the Rio+20 (Earth Summit 2012) 192 countries ultimately reached a consensus in 2012. 193 Parties to the Paris Convention recently agreed at the Paris Conference in July 2015 to set goals for reducing global warming to less than 2°C, in line with pre-industrial levels, and also to minimize temperature increases to 1.5°C. The deal plans for a half-life of zero GGE between 2030 and 2050 (Cullen et al., 2011).

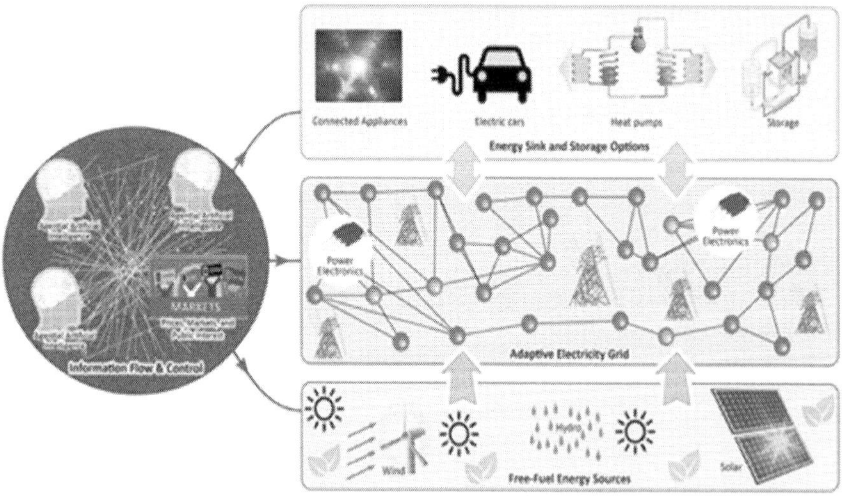

Figure 1.3: The electrification of energy: Long-term trends

Source: *https://www.cambridge.org/core/journals/mrs-energy-and-sustainability/article/electrification-of-energy-longterm-trends-and-opportunities/10440EBB31859DBE848C28AC70AAE8FB*

1.6.4. Solving Energy Problems

Researchers, technologists and decision-makers have been inspired to revisit the energy policies and priorities of the energy and sustainability issues discussed above. It is now known that electricity source decarbonization has to be achieved very quickly to achieve zero net GGE. Smart grids would need to connect countries in such a way that surplus electricity is accessible and distributed efficiently (Chowdhury et al., 2020). Around the same time, more energy storage systems are being designed to save extra energy in times of energy shortages or to enable the use of stored energy in other contexts. However, in order to minimize electricity demands, the use of resources must be much more effective.

1.6.5. Increasing Energy Efficiency and Reducing Energy Intensity

Higher energy consumption can be accomplished by increasing energy usage. This can be accomplished by improving the construction of machines, generators and, in particular, homes, which make electricity one of the highest levels of consumption. Light, heating and air-conditioning pollution conservation, more natural energies being used, urban water and electricity recovery, etc. should be used to boost energy efficiency in buildings with near zero energy (NZEB) or zero energy (ZEB). It is also important to better plan communities and mobility within these regions in order to minimize the usage of individual vehicles, thus reducing not just the GGE involved, but also the share of energy use in transport. Product growth must be matched differently with a reduction in the distance from the processing facility to the place of use, thus reducing transport resources and increasing the quality of production. This reduction in energy usage would also lead to lower energy consumption, i.e. a lower energy intensity in commodities. Improved waste collection technologies for the recycling of capital and energy storage are now mandatory in order to improve energy production and minimize energy intensity.

1.6.6 Increasing the Sustainability of Energy Systems

The prospect of constructing bio-refineries that are close to a petrochemical complex but rely on the use of renewable raw materials and energy was one of the suggestions that was probably deemed more realistic in the medium to long term. They are a smart option from a conservation point of view, as they allow renewable energy to be complemented by non-renewable energy, thereby reducing the net environmental impact of the products on the life cycle. In comparison to current energy sources, in particular fossil fuels, it will also be possible to diversify future RES applications in order to improve efficiency, longevity and economic potential.

This is definitely a rapidly evolving world in which a lot of investment in R&D and expanded production capacity is being made. While it is currently facing major obstacles, it is agreed that it will be a key component of potential energy networks in the longer term for all related stakeholders in the future. It is also understood that the ideals of economic development, application and operation should be taken into account as far as possible. This has also been achieved with some RES, in particular biofuels, in which several companies and national or international agencies also have approval systems in place. Such forms of RES, in particular biofuels, can also be found in the literature for sustainability evaluations. However, questions remain unanswered about their effect on sustainability in other RES, and more objective methodologies are required for their assessment (De Andrade et al., 2015). In the other hand, there is an immense energy reserve inside waste and bio-waste. Specifically, to facilitate the sustainable reuse of resources and to recycle all types of material, waste and bio-waste in the event that it is no longer possible to ultimately become oil, thus reducing environmental dangers and energy losses. Reliable, efficient waste management systems are expected.

1.7 PRODUCTION OF ENERGY STORAGE SYSTEMS

Energy storage systems are important for maximizing the efficiency and utilization of energy from the current RES and/or potential RES. In particular, the option of energy storage and/or technology that is more efficient will greatly alleviate or even overcome many of the more critical issues and problems that affect current RESs. For example, improving the management of demand and supply flow, reducing climate-related variability in energy output and enabling energy generation to be used in other processes. The problem is critical to balance demand and supply, as electricity from wind and photovoltaic energy systems is an important component of renewable energy generation. The growing production of renewable energy also relies on photovoltaic or wind energy, although other options, including hydroelectricity, are still ongoing with low potential growth capacity and others, such as wave power. This pattern has also been noticed internationally, with much of the spending on green energies, such as wind and solar, in particular photovoltaic energy. In certain cases, Ses RES causes a surplus of power that cannot be used due to current storage issues. Returning to the RES would have significant short-term and medium-term consequences, especially at the economic level, if the losses of renewable energy were reduced or the supply/demand balance improved. Energy management is important, thus reducing the influence of external forces and, in particular, of temperature situations, in order to ensure that the supply and demand balance is as near as possible.

It is even more complicated because there are fragmented RESs, particularly wind and solar, which pose major challenges to the supply of grids and energy storage systems. The US Department of Energy's Energy Storage Technology Initiative provides a strong project database. It can be concluded from the list that currently most of the power plants installed or under construction are pumped hydroelectricity. While current dams and other facilities (especially power grids) are always available, there is limited storage space and, due to the lack of climate or suitable areas, there is little choice in many parts of the world (Delmastro & Gargiulo, 2020). Moreover, the operation of these systems that have significant environmental implications and their viability is frequently not properly measured, although environmental impact analyzes are essential for these types of projects. At present, a wide variety of energy storage systems, such as cars with a lead acid battery that stores the energy needed to fuel a generator, are used in a number of contexts and circumstances, including our daily lives. A variety of methods are available that concentrate on distinct physical and/or chemical phenomena. For example: mechanical storage of energy as air and hydropower storage, thermal storage as phase shift or thermal storage underwater, electrical storage as batteries, chemical storage, hydrogen, etc. Examples included: The various methods rely on power, energy density, timeframe within which energy can be stored, and possible uses of stored energy, among other things. In order to have a significant impact on the global energy balance of energy storage systems for use in RES, large quantities of energy storage systems should be capable of storing energy at least within a reasonable time in energy supply-unsatisfactory environments and should be sufficiently scalable to be used in a variety of settings or systems. Chemical methods for over-storage of renewable energy are increasingly seen as a good choice between possible solutions, in particular with a view to increasing market penetration of renewable energy. Here are some of the main advantages:

It has been an energy storage facility for a long time, and energy quality is not lost in time if properly managed.

- They can be driven and transported by electricity.
- There is also a clear awareness of many facets of chemical packaging and technical expertise, including the processing, storage and supply of energy, is applicable at all stages of the life cycle.

In addition, surpluses of renewable energy could be needed in other settings depending on the form of energy storage. For example, it will be best to provide current and/or potential RES, improve efficiency and overall performance, or include capital in existing processes. The potential use of the energy generated must be diversified and certain methods of energy storage, in particular methods that focus on chemical materials, may be used to increase the economic value of the current and future RES. There are two examples of electricity generated by water electrolysis in wind turbines that can act as transport fuel or methane produced by reactions in the chemical industry that can be used. In both cases, the existing gas delivery system, whether or not combined with natural gas, can be used for both storage and/or supply, for which there is now a great deal of expertise and substantial infrastructure. This enhances the existing and/or potential viability of RESs, facilitates the integration of RESs and enables surplus energy to be used as fuel in other contexts and industries, especially in the transport sector. It is especially necessary to use storage energy for the chemical and process industries in order to generate not only other fuels, but also higher chemical feedstock's or chemicals such as the gasification of Fisher's Tropsch synthesis, but also higher productivity feedstock's such as microalgae and residual biomass residues (Edelenbosch et al., 2020). The incorporation of new RENs into existing industrial systems would allow the transition to a greener and more bio-based economy easily consistent with the concept of bio-refinery. Although this alternative is obviously of interest and the numerous developments and/or solutions suggested, further work needs to be done to better expand and reach the full potential of renewable energy technology.

Chapter 2

Usage Of Energy in The Production of Agriculture

CONTENTS

2.1 INTRODUCTION

Present day cultivating needs current energy - both have a cozy connection. Horticulture is the primary field in the development of the economy for some agricultural nations. The fundamental drivers of worldwide destitution decline are expanded effectiveness and the modernization of farm frameworks. Energy assumes a significant part in that. The contribution of energy into present day and practical horticulture is a critical factor in moving from resource cultivating to food security, added an incentive in rural regions and the advancement of new agrarian business sectors. Frequently, energy benefits that help a gainful activity, for example by flooding (siphoning) or treating (cooling) or handling after reaping, can be upheld by sustainable power innovations and crossover frameworks (squeezing, processing, drying). The mechanical energy necessities are likewise significant in the farming creation measure, covering human and animal work and motorization, siphoning and different tasks just as by implication treat and agrochemical creation.

Figure 2.1: As an essential contribution to development, farming needs energy. For machinery and equipment, for heating or cooling buildings and for illumination on the site, and in fertilizers and chemicals generated off-farm, farming requires electricity either as fuel or electricity.

Source: https://urbasolar.com/energy-production-in-an-agricultural-setting/

As an assembling area, horticulture requires energy as a fundamental contribution to improvement. American cultivating – for the animal and crop items – has become an ever-increasing number of motorized and needs ideal supplies of energy at specific periods of the creation cycle to accomplish full profitability. This section highlights certain important points:

- Farming depends on an opportune stock of energy, yet energy utilization is diminishing.
- Farming in the United States uses energy directly in manures and off-field synthetic compounds to create fuel or power for planting.
- Contingent upon area, request and exercises, the portion of power in farming creation costs is profoundly a cause.
- Direct energy costs are a significant piece of the general creation costs in many exercises and creation measures on the rural level, however moderately little.
- Complete energy expenses can assume a lot larger part in farm overall gains, specifically for huge field crops, when joined with roundabout energy consumptions.
- Changes in the cost of energy have repercussions for yield and activity blend farming decisions, development methods, and water system techniques and post-reap methodologies.

This paper gives a viewpoint on the energy-agribusiness relationship in the United States. The primary section gives setting data on current and authentic utilize and how this mix into the public picture of energy use in the United States farming area. The part of Energy in the general expense design of horticulture is itemized for both present and time conditions. At long last, the investigation talks about the scope of exercises and locales in horticulture's example of energy-use (Elghawi & El-Osta, 2015).

At the farm stage, either direct or indirect energy usage is listed. Direct energy use in agriculture consists mainly of oil-based fuels used in the service of trucks, pickups, and vehicles, as well as output and input equipment, for preparation of harvests, crown plants, and fields, transport, and chemical application. The power plants and irrigation equipment also use electricity, liquid propane, and natural gas. Electricity is primarily used in homes and barns for illumination, heating, cooling. Dairies will need electricity for sanitary, cooling and working milking systems to supply hot water. Besides, for a wide range of farms, oils and oils are required. In the sector of the production of manures and pesticides, aberrant energy is burned through. Because of figuring issues energy isn't essential for the USDA's idea of aberrant energy for creating other ranch inputs, for example, ranches hardware and gear.

2.2 FARM EQUIPMENT ENERGY CONSUMPTION

Each interaction of farming includes the utilization of petroleum derivatives consuming gear and apparatus. On the off chance that we utilize such powers, we will raise the temperature of the Planet and add greenhouse gasses noticeable all around, for example, nitrous oxide, methane, and carbon dioxide. While a tractor might be the primary piece of gear to be recollected, other energy-using machinery are not as determined. We by and large do not compare butchering animals with the use of water, for instance, anyway we should. One animal can burn-through up to 132 gallons of water when butchered.

This is basic on the grounds that the water does not simply turn up – it must be siphoned all through the abattoir. These siphons work on non-renewable energy sources, power. The fact that around 55 billion creatures are murdered each year is unbelievable (Ersel et al., 2004). Although ranchers' interest in meat cannot be managed, there is yet an approach to decrease reliance on petroleum products in farming. Specifically, the soil disintegration is diminished by ranchers who increment no-work land - for example by not developing discharges prior to planting seeds - and by and large diesel utilization is diminishing. Truth be told, if ranchers aggregately multiplied the real estate that is no-till, it could set aside up to 217 million gallons of diesel fuel every year.

2.3 CONSUMPTION OF ENERGY IN TRANSPORTING, PACKAGING, AND PROCESSING FOOD

It is just essential for the condition to create food, regardless of whether by the development of or raising cows for slaughtering. The way the food enters the foundation additionally means for the dependence of farming on petroleum products. Albeit the fuel utilized for the transportation of food is monstrous, we additionally have a waste field: packing even though we simply eat neighborhood dairy and produce. Much the same as a ton of non-renewable energy source is important for running the water siphoning equipment into disinfection of slaughterhouses, it requires a lot of petroleum product for the machines creating the vital packing to move food, regardless of whether across the world or right down the road.

By merely using refurbished sacks for the transport of products the farmers will reduce their carbon footprint. Anything we buy in supermarkets is brought in – be it from all over the country or across countries – and in general, growers use these plastic shopping bags to ship items such as bulk watermelons and maize or to ship crop (Esen, 2016). Farmers used to burn or put those bags in the garbage. However, it is environmentally effective to reuse the packages and the additional benefits are the reduction of the quantity of waste generated every year at waste disposal.

2.4 CONSUMPTION OF ENERGY IN THE PRODUCTION OF FERTILIZER

For photosynthesis, nitrogen is required, therefore, the soil must contain nitrogen in exchange for plants to develop Higher nitrogen in the soil would increase the harvest. Applying fertilizer is the best way to raise the nitrogen level. However, we require equivalent quantities of manure if we have large quantities of food production.

Figure 2.2: The secure availability of power for Africa's crop yields is one of several essential needs for substantial development. Connection to electricity and fuels for farming or crop production is restricted and expensive for farmers in most African countries. (Source: http://large.stanford.edu/courses/2018/ph240/lascurain1/)

Synthetic compost gives sufficient soil supplement to create more harvests on a similar measure of soil. Consistently, universally, more than 100 million tons are utilized. It costs a ton of compost. Engineered composts utilize non-sustainable sources – petroleum products specifically.

Engineered compost cannot be delivered without petroleum derivatives. Consuming petroleum derivatives prompts the increment in the temperature of the earth, as expressed prior. Truth be told, it is 0.5 to 2% of all out an Earth-wide temperature boost emanation that is delivered and spread by composts. Fundamentally, we have a very appeal for engineered compost, which causes genuine natural damage.

A segment of the immediate, backhanded and solar- powered energy utilized in cultivating has been changed over into substance energy stockpiling. The rest is wasted energy. Because of direct energy, the warmth of the hardware utilized is squandered, backhanded energy is squandered when manures and pesticides miss the objective. For corn specifically, the rough energy output: Input proportion is 1.34, on the off chance that we accept direct in addition to circuitous energy as information sources and convert it to ethanol as yield. Sun-based force is prevalently lost as warmth is illuminated via water, soil, air and so forth (Ghalehkhondabi et al., 2017). For instance, in maize and soybean, only 15% and 7%, because of photosynthesis, of the net solar radiation stored in heat, canopy, and soil. It is not beneficial to dissipate much of the portion stored under soil and canopy heat and does not produce biomass. This would considerably reduce the energy output: input ratio considering solar energy.

2.5 AGRICULTURE AS A PROPORTION OF THE ENERGY USE IN UNITED STATES

Direct use of energy: Around 1.1 quadrillion Btu of the absolute direct energy is burned through in 2002 in the U.S. horticultural industry (counting both animals and harvests), which represents a little more than 1 percent of all out U.S. energy utilization of 98 quadrillion Btu in 2002. Conversely, 31.4 Btu quadrillion (32 percent) was utilized for the non-farming piece of the mechanical area while 26.5 Btu quadrillion was utilized in the vehicle area (27 percent). Because of its restricted offer, in the United States rural area's immediate energy utilization are probably not going to fundamentally affect America's general market interest in energy. However, shifts in energy supply and demand may have important consequences for U.S. agriculture's viability and the balance of production and production activities within the agricultural industry.

Indirect use of energy: In contrast with direct oil, agribusiness is significantly higher in the extent of two fundamental roundabout employments of energy, manures, and pesticides. Farming in 2002 addressed roughly 56 percent of the all-out utilization of nitrogen in the U.S. (13 million out of approximately 21.4 million metric tons). The principal manure utilized in the U.S. horticultural area is a nitrogen compost. There was not promptly accessible information on the farm's extent of phosphorous and potash compost (Grunewald & Diakonova, 2018). Besides, in 2001, US farming was assessed in the Environmental Protection Agency as containing 67 percent of pesticide spending in the United States.

Even though agriculture is the lowest source of energy directly using natural gas, its significance is enhanced by an indirect connection to feed stocks, especially nitrogen fertilizer. Natural gas is the key component for many nitric fertilizers and accounts for up to 90 percent of the cost of the development of anhydrous ammonia. In the processing of phosphates (15% to 30% of manufacturing cost) and potash (15%), however, natural gas is also a major cost factor to a smaller degree. If the components of composts and pesticides are converted into flammable gas and oil, the joint immediate and roundabout utilization of gas would be more than 26% of complete rural energy utilization.

2.6 USE OF ENERGY BY SOURCE IN THE AGRICULTURAL SECTOR

Of the Btu reported 1, 7 million, 65 percent (1, 1 Quadrillion Btu) of total energy used in agriculture, compared with 3.5 percent (0, 6 Quadrillion Btu) used in indirect energy consumption (natural gas, LP gas, diesel, gasoline, electricity) (pesticides and fertilizers). The utilization of farming's assets crested at 2.4 Btu in 1978) (Goiri et al., 2015). The peaks in oil costs in the last part of the 1970s and mid-1980s prompted an ascent in energy creation for the ranch area. The immediate energy use of farming has diminished by 26% since the last part of the 1970s, while energy for compost and pesticide creation has diminished by 31%. The progress from gas to eco-friendly diesel engines, (which are less effective), and the change to bigger multifunctional machines and the improvement of new techniques to dry and water crops, prompted this lessening in energy utilization.

Energy organization has moved over the long run. Over the most recent forty years, the general portion of fuel as the source of farm energy dropped drastically from a portion of 42% in 1966 to around 9% in 2002. There was likewise a drop in the immediate utilization of flammable gas and LP gas from joined 16% to 9%. Then again, diesel fuel and power both expanded altogether from 13% and 6% in 1965 to 27% and 21% in 2002 individually. The change from petroleum apparatus to diesel hardware is the basic reason for the expansion in diesel and the diminishing in petroleum. As far as the miles a gallon and miles a Btu, diesel is a lot higher than petroleum. The all out drop in the complete direct utilization of energy is likewise the aftereffect of a huge drop in the load of rural hardware, gear and engine vehicles that happened as all out-farm apparatus stocks crested in 1979. This drop is additionally reflected in a decrease altogether direct energy utilization. Further, between 1980 and mid-1990, capital devaluation outperformed capital consumption.

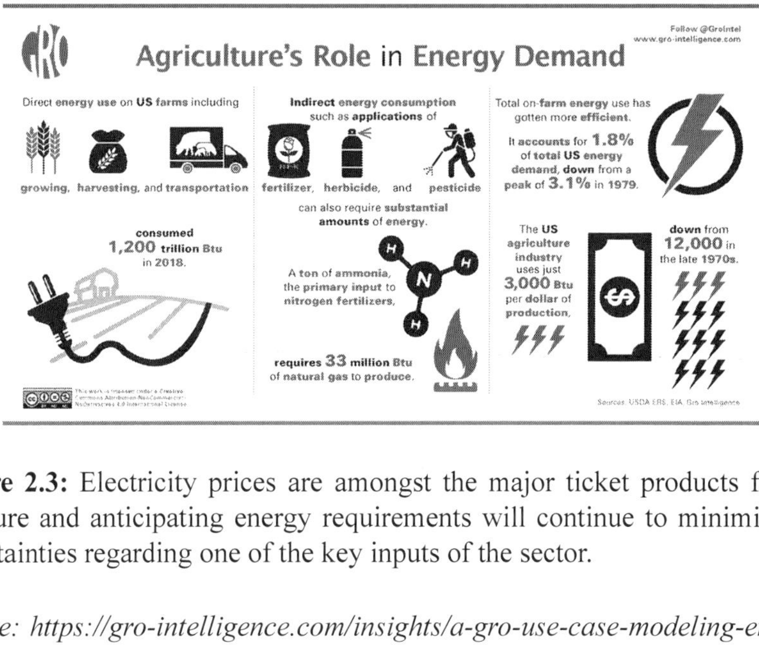

Figure 2.3: Electricity prices are amongst the major ticket products for agriculture and anticipating energy requirements will continue to minimize the uncertainties regarding one of the key inputs of the sector.

Source: https://gro-intelligence.com/insights/a-gro-use-case-modeling-energy-needs-in-agriculture

The lessening in capital was brought about by a few variables, including initially, the expanded machine adequacy and change from customary laying practices to diminished or no-till rehearses (which permitted soil to be worked a few times prior to planting) (which require less soil and, accordingly, less fuel utilization). Furthermore, soil dampness and supplements (decrease of the requirement for business composts) have been preserved in a way that has assisted with broadening the valuable existences of farm haulers and apparatus.

Since 1974, the use of fertilizers and pesticides has been disconnected from a cumulative 25% share in 1965 to marginally more than 46% share in 1986, and then declined to 35% by 2002. The result has been a disjointed pattern for U.S. agriculture. Increased use of accurate farming (i.e., computerized equipment for specific inputs, like pesticides, fertilizers, and other fertilizer products), conservation labor, and maintenance of crop residues attributed all to lower fertilizer volumes despite compromising profits (Hamlet et al., 2010). The lower pesticide quantity was also attributed by seeding on genetic engineering plantation plants such as Bt corn and Bt Cotton, which need less pesticide usage. Furthermore, enhanced pesticide formulations and increased use of crop surveillance services have helped minimize pesticide quantities while retaining the pest control standard.

Farm energy efficiency gains. After the late 1970s, significant reductions have not been at the cost of lower outcomes in the agricultural sector using indirect and direct energy sources. There seem to be dramatic increases in energy production in agriculture. The gains are calculated by a sharp decreasing energy usage for both indirect and direct energy categories per unit of performance indices. Direct energy consumption (DEU) per production unit has been almost constantly decreasing since 1980, while output total in farming has steadily increased. Each unit of output has also been decreased by indirect energy consumption (IEU) and has more variation than direct energy usage. In the 1990s, both indirect and direct energy consumption per production unit seemed to be plateauing.

2.7 THE SHARE OF ENERGY IN THE COSTS OF AGRICULTURAL PRODUCTION

Manufacturers are slowly acquiring choices on energy value shifts, yet most of ascends in energy costs keep on bringing about lower ranch pay for the time being. During the time frame 2000-2003, normal yearly creation costs for U.S. ranchers added up to around \$194 billion. Practically 15%, or an expected \$28.8 billion, of that whole was burnt through on effort. The extent of energy use per annum in horticultural creation shifts with yearly changes in planted sections of land, harvest, and related energy costs.

The cost of direct energy: In horticulture creation interest for refined oil-based goods including LP gas, fuel, and diesel is resolved essentially by the quantity of hectares developed and gathered and the climate, just as the diverse energy sorts of costs. As most of the energy utilized in America (and on the planet) comes from oil sources, for example, flammable gas, LP gas or diesel, or gas, costs will in general move together. Their costs will in general change. This declines the viability of fuel changing to energy reserve funds. Direct electrical expenses (for example for energy and oil supplies) went extensively over the 1960s and 1970s, including 4 percent to 8 percent of ranch costs. Since the mid-1990s, nonetheless, the portion of direct energy in the all-out expenses of horticulture was about 5% (Hossain et al., 2017). The portion of creation expenses of power went down to 1.9% by 1989 from about 0.7% during the 1970s and has stayed consistent as specialized proficiency drives in energy use generally made up at greater expenses.

Indirect costs of energy: In the previous 40 years, there has been a significant fluctuation in roundabout energy costs (for composts and pesticides) going from 8 to 12 percent of all out horticultural yield costs. In 1974, roundabout energy costs encountered a sharp increment as manure costs hopped. The expense shared move was the most noteworthy. U.S. compost costs were frozen at the manufacturer's level in 1971 by the USDA's Economic Stabilization Policy. These value checks were abrogated on 25 October 1973 and the U.S. compost costs and costs rose quickly. Since 1996, the portion of circuitous energy in the general expense of cultivating has diminished to around 9% in 2003.

Costs of agricultural chemicals: Agrarian synthetic spending incorporates most pesticides. In three essential gatherings — fungicides, insect poisons, and herbicides, pesticides are generally separate. Another major horticultural synthetic class includes defoliants, utilized chiefly for cotton in the United States. The share of agricultural production costs in the pesticide sector grew dramatically from below 1% after 1960 to approximately 5% in 1998. The rise in costs that took place before 1980 was due to both higher overall consumption and raised per unit cost, although between 1980 and 1998, the rise in costs was primarily due to higher prices per unit charged. Total stock of active ingredients of agricultural chemical products added to plants grew gradually from the early 1960s to around 1980. Quality expansions in pesticide mixes, be that as it may, have been supported through the 1990s, their capacity to target chosen target pests, and ranchers' ability to more likely objective pesticide applications. Since 1980, these and different upgrades in quality have diminished utilization rates, however, prompted the ascent in per unit costs charged during the mid-1990s (Huang et al., 2011).

Costs of fertilizer production: In 2002, there was around five percent of rural yield consumption in compost use. In any case, they were the greatest use on farming with a 34 percent portion of the general energy consumption of 28 billion dollars in 2002. In the actual year, manure was likewise the greatest single source (estimated in Btu's) of agrarian assets, with 29% of its offer. U.S. farming has used almost 20 million metric tons of total fertilizer since 1991. Of this number, the largest portion of the nitrogen-based fertilizers consisted of 56%, in comparison to 21% for phosphate and 24% for potash. Demand for fertilizer depends on different factors, including both output and input relative prices, crop rotations, climate, and fertility and soil type. Many plants, if not most, have routine use of the trade fertilizer in the United States. Fertilizers provide nutrients that increase both crop and plant growth.

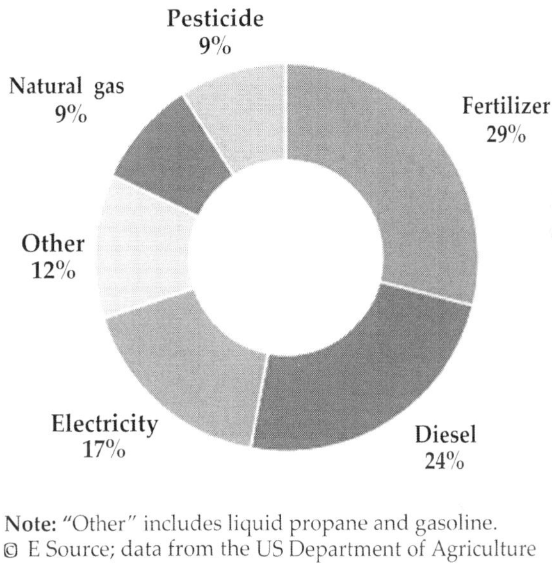

Figure 2.4: The usage of power is frequently divided into direct and indirect uses. Direct use of energy is a measurement of yearly utilization of water, electricity, oil, wood, gasoline, telephone, and online services. Indirect energy usage is the incarnated energy utilized to produce some fertilizer and pesticide products. (Source: https://esource.bizenergyadvisor.com/article/agriculture)

Most US nitrogen imports are traditionally imported from Canada; (accounting for about 40 percent of total imports since 1989). But nitrogen imports from countries such as Saudi Arabia, Qatar, Kuwait, Egypt, Bahrein and Venezuela, Tobago and Trinidad, Ukraine, Norway, the Netherlands, Romania, Poland, Russia, China, and Bulgaria have increased since 2000. The United States has also increased its import share. The prices of fertilizers are linked to the prices of Natural gases: The creation of US manures is rigidly connected to energy supply, especially flammable gas. In the creation of anhydrous alkali, gaseous petrol is the principal segment. Anhydrous ammonium is utilized for the assembling of most other nitrogen manures as the crucial structure block (e.g., nitrogen, ammonium nitrate, urea, and nitrogen arrangements). The handling of these other nitrogen manures from anhydrous alkali additionally utilizes common gaseous petrol as a cycle gas. In summary, petroleum gas represents 75% to 90% of nitrogen manure creation costs (Liu et al., 2019). Thus, flammable gas represents 75% to 90% of nitrogen compost creation costs. Flammable gas likewise gives a significant type of revenue for the creation of di-ammonium or mono-ammonium phosphates (15% to 30% of creation expenses) and potash (representing however much 15 percent of the creation cost).

Since fertilizer prices are intricately linked with the prices of natural gas - through anhydrous ammonia, these prices shift in tandem with natural gas prices due to anhydrous ammonia and the prices of other nitrogen fertilizers are in turn less intimately linked to gas than the prices of nitrogen fertilizers due to phosphate and potash prices. Two possible solutions are encouraged to higher fertilizer prices: (1) lower fertilizer applicable in the current crop mix; or (2) lesser fertilizer dependence planting and output. While applications of nitrogen fertilizers in many fruit and vegetable plants appear to be higher, field cultivations in considerably larger fields are plants. As a result, the overall use of fertilizers is most prevalent in crops planted in the largest region – maize and wheat that lag well behind sorghum, cotton, and rice.

2.8 USE OF ENERGY BY THE AGRICULTURAL ACTIVITY

Both creation and territory contrast altogether in creation consumptions and in the overall meaning of the energy costs. While a few unique kinds of farms work, practically completely motorized fields and commercialization and the board exercises incorporate apparatus (for example work vehicles and collectors) just as cars and trucks that depend on oil. Grain dryers and water system frameworks are normally more adaptable on the grounds that they are provided with oil, flammable gas, and power, while power is a significant lighting, warming and refrigeration power source in houses, stables, and different farms.

Registration measurements show that energy uses on horticulture were $24 billion out of 2002, made in yields out of $18.4 billion and animal production of $5.7 billion. Almost 14 percent of the US' net agrarian yield costs represented energy costs. Regarding the portion of energy costs in each significant creation activity, 23% of yield creation costs is because of energy costs and simply 6% to animals' expenses. The more energy is liable for the portion of in general creation costs, the more energy costs or supply changes are vulnerable to a creation movement (Liu et al., 2013).

Significant crops in fields: Major agricultural production historically needs many crossings, either by pulling a tractor or using a specialized machine that can perform one or more of these functions, to prepare, plant, expand, fertilize and chemical applications or harvest. The fuel consumption depends on the efficiency and the number of passes (determined by the used tillage practices) of the machine concerned, as well as the size of the area. Indirect use of energy in fertilizers and pesticides differs between regions and crops, depending on the production methods, soil, and temperature.

An enormous extent of the yield of US ranch crops is flooded each year which needs extra energy to run the siphoning framework. Around 55.3 million sections of land or just about 13 percent of 434.2 million sections of land of cropland were watered in 2002 for any yield zone, rummage, vegetable, and trees. The utilization of the water system differs quite a long time after a year, contingent upon the states of the soil and climate. In 1997, for instance, practically 16% (67.8 million hectares) of the all-out developed land was flooded for some 425.2 million sections of land.

The water usage can also differ dramatically depending on crops grown — only approximately 6% of agricultural output was irrigated by 100% of rice crops in 1997. When picked, most plants on the field need extra energy-related treatments on the farm until they are sold. Crops that are collected with a high humidity component are usually dried for storage and processing purposes. Other crops, including tobacco and cotton, need additional energy expenditure. To isolate the lint from foreign matter and seeds, cotton must be ginned. Tobacco must be cured — a heated and dried method for the production and retention of usable tobacco purity, aroma, and flavor — until cigarettes and certain items can be processed (Liuyijuna, 2011). The extremely aggregated group of major crops was, as per the 2002 Agricultural Census, the greatest agriculture consumer, at both $13.6 billion in overall spending and with a 27% share in manufacturing costs. In comparison, the energy prices of big crops on the ground' constituted 29 percent of the U.S. agriculture overall energy costs.

Fruits and Vegetables: Manufacturing activity for leafy foods differs generally, from profoundly automated creation with low work contribution to work escalated creation with low mechanical levels. The water system is additionally utilized in vegetables and organic product creation, and the ordinary segments of the creation cycle comprise synthetic substances and composts. Field warmers or windmills are utilized to relieve the potential impacts of cold temperatures in some citrus and other organic product developing locales. In 2002, the energy expenses of "vegetables and melon" added up to 22% of general manufacturing costs of $2.0 billion. In an assessment, energy costs "leafy foods and nuts" of 1,7 billion dollars represented 17% of the general expenses of creation. Greenhouse, nursery, and flower farming. Temperature management, plant diseases and control pests, fertilization, and timely irrigation activities, among other things, constitute less than 10 percent of total costs in terms of productivity in greenhouses, nurseries, and floricultural fields.

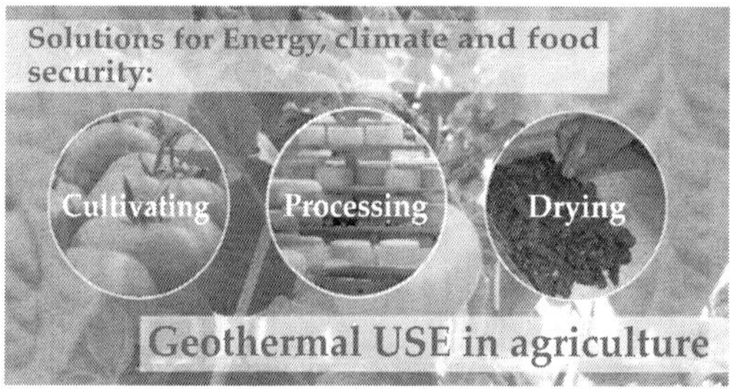

Figure 2.5: Geothermal heat is accessible worldwide at low to medium temperatures and structures that make their application easy and convenient to manage. Geothermal systems are centrally built and at reasonable rates include heating and cooling. It generates direct and indirect jobs across the entire supply chain. (Source: https://www.thinkgeoenergy.com/egec-geothermal-energy-use-in-agriculture/)

Ranching of beef cattle: The key energy use exercises engaged with cow-calf tasks and other steers munching exercises are field and promoting exercises. The administration of fields incorporates weed control, preparation, and water system, in numerous areas. In 2002, energy consumptions comprised *roughly* 12% of the general farming of steers meat. The farming record for a generous (practically 12%) portion of homegrown rural energy utilization, including over 15% of the costs of fuel and 10% of the costs of manure utilized by the United States farming in 2002, notwithstanding its low portion of in general creation costs. The huge number of animals offered to feeding stuff or butcher houses every year is clarified by the enormous measure of energy that comes from the hamburger farming industry in the

USA (almost 420 million sections of land) (in 2003, 18.4 million head of steers and calves were butchered, while as of 1st July 2004, 11.8 million head were on feed).

Production of aquaculture: The improvement of hydroponics comprises of the fish cultivating of huge fish species, for example, salmon, catfish, and so forth and of mustards and shrimps. Species and assembling strategies contrast in energy utilization, yet it can incorporate both fingerling tanks and concentrated rearing tanks. The vital users of power are temperature and water guidelines just as lighting. In the 2002 farming registration, hydroponics is recorded as "other animal horticulture exercises." Together, the energy cost of this composite gathering was $445 million, representing just about 8 percent of complete manufacturing use (Majidi, 2014).

Milk and dairy cattle production: The draining tasks need heated water for sterilization purposes, milk cooling, and power to work draining frameworks. Field the executives, taking care of and advertising both straightforwardly and in a roundabout way burn-through energy. Complete energy expenses of $1.2 billion were under 7 percent of their all-out creation consumption for milk and dairy items in 2002.

Feedlots for cattle: Feedlots use resources for feeding animals and water, for treatment of animal residues, and for the selling of animals for plant packing and other abattoirs. In 2002, nevertheless, feedlot power bills of 2.3 billion dollars represented less than 3% of the overall cost of production. Cost spending was dominated by buying feeding-stuffs and feed.

Pork and Hog production: Some hog manufacturers use some forms of containment processing with advanced environmental facilities. Hog barns, nurseries, and central farrowing require watering, feeding, cooling, and heating systems for electricity. For pork and hog processing, a total cost of 526 million dollars accounted for less than 5% of the total cost of production of 2002.

Production of eggs and poultry: Most poultry and egg laying is completed in expert structures, like the case with hog creation. Chickens don't have a lot of room until they have appropriate food, ventilation, and clean new water all nonstop. Chicken need not bother with much room. As a result, the watering, taking care of, cooling, warming, and lighting frameworks of poultry reproducing, and creation offices are required. For poultry and egg creation in 2002, absolute energy expenses of 534 million dollars addressed 3 percent of their general creation consumption (Mortazavi et al., 2019).

2.9 PREPARATION, HARVESTING, CULTIVATION AND THRESHING OF LAND

Human labor, draught livestock or motor-powered machinery are the power required for agricultural production. In any agricultural system that incorporates tools, equipment, and machines to enhance agricultural labor and land efficiency, mechanisms are a crucial input. The technical stages of mechanization have been commonly categorized as hand tool engineering, drawing animal engineering and mechanical energy technology, based on energy sources. Most of the direct energy input is generated by human and animal power in sustainable agriculture. Based on the accessible physical energy, farming in developing nations depends highly on farmers' physiological capacity with sometimes small outputs. A fit person uses approximately 250-300 W of energy, according to the environment, with 10-30 minutes an hour remaining. With a combined power output of 75 W, the power conversion efficiency is only 25 percent. Human skill practices include seeding, weaving, sprinkling, and collecting, as well as the use of hand-held instruments. Most agricultural products, such as those cultivated in increasingly developed nations, are collected manually. Mechanical power for digging roots and tuber plants requires the energy needs for harvest. The International Centre for Tropical Agriculture (CIAT) implemented a basic tool for manioc harvesting from 15 to 40 days (based on the type of soils and produce) to Thai farmers and reduced the number of days to just 5-10 days. This tool requires USD 2,50, is generated by local producers, and could be utilized by growers jointly. Collecting at a lower air level earlier in the mornings also allows minimize energy demands and cooling prices.

Figure 2.6: The usage of equipment (e.g., farms with farm machinery) and the heating of stocks and Greenhouses absorbs energy directly by agriculture. In addition, for the development of agricultural goods, farming equipment and houses, farming is indirectly using electricity.

Source: https://northafricapost.com/19236-morocco-invests-2-3-billion-dirhams-boost-renewable-energy-use-agriculture.htmlz

Activities like threshing, elevating, preparing the soil, and plowing need less abilities, but a greater amount of energy and are therefore guided primarily by animals of draught, such as bovines, camels, donkeys, and horses. Power range for 4 hours a day from 200 W for an ass to over 500 W for a buffalo.

Draught livestock energy is an inexpensive and reliable power source for small - scale farmers and is also regarded a major resource of energy for diversion and growth, for instance in terms of transportation and non-farm jobs (Moran et al., 2014). Works on animal productivity that can be enhanced with equipment modernization, improved breeding, and livestock farming, feeding and veterinary care should be included as part of initiatives to maximize the use of drawn animal power.

In Sub-Saharan Africa, the primary interest tops in cultivating cycles are around 65% of the energy required for land arrangement and ensuing weeding. An examination uncovers that only 1.5 hectares each year can be filled in a run of the mill ranch family subordinate only on human strength. On the off chance that draught animal power is open, this will rise to 4 hectares and more than 8 hectares if farm hauler force can be gotten. The change to expanded horticultural motorization happens through work vehicles and collectors fueled by petroleum or diesel.

The two-wheel work vehicles are normally utilized in cultivating with a scope of culturing connections, sheeting, line estate, reaping and sifting. Force turners (12-15 hp), because of their adaptability in Bangladesh, are utilized for about 70% of farm laborers. Big haulers and tuber crops, for example, potato diggers, should be taken care of 0.57 gallons of diesel fuel per ton (1.96 liters/MT). Fuel per ton of item. The utilization of assets generally relies upon the speed of exercises. Since laying in arable systems often requires the highest energy consumption and leads to soil degradation, mechanical laying decrease is encouraged by a conservation agriculture approach (CA). Generally, for a smallholder with a typical land holding of up to 5 hectares, a four-wheel tractor is not cost-effective. The SSA countries have not succeeded in increasing the productivity of the farm through government-run tractor hire systems. On the other hand, there is potential to increase the concept of the rental market for private and controlled tractors in the future.

Aside from direct energy interest in agribusiness, roundabout energy inputs, all of which require energy for their preparing, dissemination and transport measures are sequestered energy in bug sprays, pesticides, herbicides, and composts. The biggest contribution of energy in farming is manure, while most energy-vital data sources are pesticides (for every kg of compound). In any case, the utilization of natural fertilizer from the agribusiness or bio fertilizer creation based on horticulture will limit the interest for these data sources, for example from biogas creation plants (at the family unit or mechanical level). The advancement of neighborhood natural manure likewise has the advantage of dodging the negative ecological impacts of mineral compost and of decreasing energy prerequisites and expenses of fuel by exploiting more limited distances to be distributable from them (Muralitharan et al., 2018).

Studies from Asia shows that guaranteed markets and horticultural costs for key yields, good government strategies and admittance to advances are the critical drivers of automation and consequently power for rural creation. India's model shows that positive government strategy and interest for ranchers has prompted expanding accessibility of reasonable and privately created agrarian instruments, lastly India is the main producer of farm haulers on the planet. The National Sustainable Agriculture Information Service of the US National Technology Center has performed an incredibly detailed analysis of energy usage in traditional and ecological cropping systems. In a study of existing research, the analysis reveals that a comparison of energy use in traditional and organic farming systems involves several challenges. Some evidence shows that organic farming is more power effective than traditional farming, but not all. In certain situations, based on the agricultural process, crop manufacturing and handling of the postproduction, organic agricultural activity can be energy consuming. The energy intensity of food production should be measured in a holistic way, which involves energy usage over the life cycle of the food commodity.

2.9.1 Irrigation

Water access is a key driver of agricultural yields and successful irrigation water management provides a significant incentive to stabilize yields. Consequently, irrigation is extremely important in the farm process and thus this chapter is especially important. The output of agricultural area is far more than twice that of irrigated agricultural land. Just 4% of the development area in Sub-Saharan Africa is irrigated compared to 39% for South Asia and 29% for Eastern Asia. With increasing water shortages and the cost of large-scale water resources, efficiency needs to be increased, existing schemes expanded and small-scale water harvesting established. Irrigation land applies, where applicable, to water supply for cultivating crops, such as the implementation of storage facilities and pumping. The energy requirement for irrigation is the energy needed to raise water by pumping from sources like reservoirs, canals, or streams, or from groundwater sources utilizing boreholes or open wells (Mukhopadhyay & Nateghi, 2017). Usually, this water is pumped into superficial canals, reservoirs, or high tanks. The energy request is controlled by expanding the volume of water to be lifted in cubic meters by the head (vertical separation from the water source to the field) (m3). Normally, siphoning energy necessities are communicated in m4 units.

Figure 2.7: Without accountability, the limitless renewable resources for heat pumps are not bound to guarantee conservation. In places where irrigation plays the most important role in crop output, particularly low rainfall levels, the necessary water is often taken from groundwater reservoirs, which are undetectably endangered by the over-use of water resources.

Source: https://energypedia.info/wiki/Sustainable_Energy_for_Pumping_and_Irrigation

There are three main sequential steps in making decisions about irrigation and energy:

- The correct water system innovation is chosen: to comprehend the water system framework, a specific setting is required.
- The privilege siphoning innovation is chosen: to comprehend the different accessible siphoning advancements.
- Choosing the correct source of energy: deciding the force sources accessible to give the necessary energy to this gadget, which are wanted and locally accessible.

The inquiry regarding which water circulation framework water system is most reasonable relies upon various elements including crops developed, climate, area, scale/ranch district, time necessities for capital access, framework cost, amount of water, neighborhood ranch laborers' specialized abilities, nearby hardware accessibility, upkeep and fix administrations and openness o Irrigation frameworks shift from top, medium water prerequisites and the general requirement for siphoning and energy stockpiling (for example supplies, reservoirs). Where little ranchers (< two hectares) are focused on, more exorbitant water system advances can be arrived at exclusively by the enrollment of cooperatives and rancher associations with fitting capital raising limit and the capacity to fund, support or supplant the innovation. For ranchers, the effectiveness and support of the picked water system framework are of vital significance (Ouedraogo et al., 2012).

There are a wide scope of water system siphoning advancements yet only one out of every odd type of water system framework is appropriate. The siphons are distinctive regarding force, size and siphoning, water source types (surface or groundwater), potential water system scale/field, cost, and specialized multifaceted nature. There are three kinds of siphons accessible: engine, manual and pressure-driven. In terms of the type of energy source used to power them, various pump technologies are also versatile. An overview of various energy sources and the description and features of each of the motorized pumps is given. An electrical pump can be operated directly where a grid link is available. However, the usage of these energy sources is dictated by prices, availability, performance, and efficiency of local electrical supplies.

Fuels for combustion engines: Mechanized siphons can be controlled by non-renewable energy sources (diesel, oil), either by power generators or by power move through a drive belt and turning shaft into the siphon. Moreover, certain submarine siphons (for example advancing cavity siphons), including cylinder siphons, work straightforwardly. The siphons are typically more exorbitant however powerful than the generators. There is likewise confined utilization of biofuels (biogas, plant-oil). There are numerous analyses being led, for instance in India with nearby olive oilseed biofuels, in the PSDA program with biogas in Kenya, and in Nepal, where Win rock International backings local area developed water system siphons with jatropha oil.

Wind energy: Mechanical and electric siphons can be fueled by wind. Corresponding, non-mechanized submarine siphons use wind-controlled mechanical siphons which need wind paces of at any rate 2.5 m/s of greatest up to 4 m/s. The limit of 1 m3/hr at profundities of 20 m or more is greatly decreased than for mechanized outward siphons. To quickly react to mechanical disappointments, mechanical siphons require nearby support and fix offices. There should be adequate wind speeds at the well site (Oughton et al., 2018). The advantage of siphoning, day, or night is that it very well may be utilized autonomously of force or fuel sources, insofar as wind is satisfactory.

One disadvantage is that they must be situated straightforwardly over the well, where nearby wind assets won't be awesome. Wind siphons are ideal just for restricted irrigable territories in breezy zones with no other force sources. Wind electric turbines change the wind's active energy into mechanical rotational energy, which drives the generator to produce power, for instance for siphoning water system water. Windmills are set for ideal wind conditions to give more site flexibility and furthermore support the advancement of power for different applications. By and large, wind turbines with an appraised yield of somewhere in the range of 1 and 10 k, we are utilized for water-siphoning applications. They are monetarily accessible a wide scope of little wind turbines with evaluated yields from several watts to 100 kilowatts, and overall, they are utilized to give power in zones where options are inaccessible or excessively expensive or not promptly accessible. Regarding hybrid power, small hybrid power systems are utilized globally, including mature technologies the combination of various sources of energy (wind diesel and solar diesel) will provide hybrids with a large and very stable supply of electricity. These hybrid systems are simple to build — there is no need for special equipment or concrete.

2.9.2 Storage and Post-harvest

Appropriate reaping and after-gather the board are the best preconditions for capacity. It is fundamental, especially for effectively operative yields like tubers, vegetables, and natural products, that item is arranging, reviewing, and pressing into compartments after reaping. Bringing down the creation time would dispose of mechanical harm and further damage among collection and use. Harvest packing units are frequently basic constructions that give the ranch laborers performing manual post-collect tasks with shade and open to working conditions. For high-worth green products, manual taking care of is suggested (Peacock et al., 2014).

Before sale on the marketplace a certain portion of the farm products must be processed as output is periodic while competition remains year-round, especially for vegetables and fruits. Safe storage centers for swift emergency assistance and reserves for local price stability safe agricultural products. In addition to storing central junctions and transport centers, it is also important to store them more locally.

Figure 2.8: The photosynthesis requires nitrogen and soils should have nitrogen to produce food. Increasing nitrogen in the soil would increase the yield. Using manure is the simplest method of growing the quantity of nitrogen. We need equivalent quantities of fertilizer therefore if we have large amounts of food development.

Source: https://energycentral.com/c/ec/how-agriculture-uses-energy

Conventional town organizations should be fortified inside the smallholder intends to accomplish specific progress to bigger and more current stockpiling frameworks to accomplish expanded creation. This calls for dry storage spaces to be controlled, especially for the post-revival and capacity of grain (rice, wheat, millet, sorghum, and mice), vegetables (peas, beans), tubers and root crops, and oleaginous seeds. Mechanical harm is regularly caused during preparing, transport, and reap of grain, and oil seeds, and harm to bugs, mold and rodents happens during storage. Energy input or expensive gear is normally utilized for such direct food sources.

Good storage practices are essential to quality and value over the entire storage cycle. A requirement to this is that crops are well washed and properly tinned until they are stored, so that live adult insects can be removed. Soil, stones, weed seeds, chaff, and strokes also must be taken away, since water is kept, and grains will dry faster when removed. The cereal must be well dried to a reasonable degree of humidity. Good drying is necessary since moldy and spoilable wet grain. You can use solar dryers – sole drying will also help you get rid of adult grain insects. In any case, since sun-drying won't slaughter the entirety of the youthful stages, for example, juveniles in grains, bug spray will regularly be needed to treat the grain, annihilating the juvenile stages when they mature and arise as grown-ups from grain. Maize is ordinarily hand-shelled. This results in better grain quality however are a sluggish interaction. A scope of precisely controlled machines can accomplish snappier and proficient shelling (Perwez et al., 2015). For both these and for safe stockpiling, the grain should be adequately dry. A decent store keeps the grain dry and cool and offers safety from rodents and birds.

Store disinfection — keeping it as spotless as could really be expected — assists with keeping the yield and the store fit as a fiddle through the whole store season. Since pests may attack the store consistently, the store and the yield should be routinely reviewed. Numerous reasonable advances and regular strategies are conceivable, from area to area to dodge post-collect damages. Little, on-ranch stockpiling to enormous and medium-sized offices range from a few hundred tons or more.

Cold stockpiling can produce high energy usage in tropical and subtropical environments. All in all, new products of the soil are at first cooled, prepared, and cold put away among the most energy-escalated food industry tasks. Cooling is by and by perhaps the most basic stages in the post-collect chain to diminish breathing rates, increment timeframe of realistic usability and shield quality for new foods grown from the ground. The decay of the products of the soil depends in enormous part on temperature during capacity, so the temperature should be brought down to a satisfactory amount, to forestall damage and to amplify the capacity time.

To preserve the quality of many high-value agricultural products, regulated temperature storage constitutes a critical factor for most perishable agricultural products and a constant cold chain is required. Cooling also provides farmers the ability to raise their profits by increasing the sale and marketing cycle when goods are available at better prices. Refrigeration in rural areas also needs to be run in developing countries without reliable electricity supplies and it is often not feasible to provide a cooling chain or storage rooms for cooling large quantities.

For the cooling of smaller numbers of produce, three technologies are available:

- Evaporative/uninvolved coolers: no energy input is fitting for temperatures between 10 and 25 °C.
- The retention coolers need temperatures under 10 degrees C; the cooling framework is fueled by the warmth as a fuel source (for example sunlight based, lamp oil energized fire). Mechanical pressure innovation is utilized in coolers electrically determined.
- Cooling depends on the solid and nonstop accessibility of power from different sources: the lattice, the diesel blower, the mechanical force created by the water turbine, sunlight-based power changed over from sun-oriented cells into batteries or sun-based energy and dissipation power.

The utilization of lamp oil or gas-filled ingestion coolers, the most well-known other options, yet those with the most negative impact on the climate, was brought into non-industrial nations in a business PV-fueled fridge stage. They are currently additional substitute techniques for putting away limited quantities of horticultural items available to be purchased, at first executed for clinical refrigeration (Rosenberg et al., 2013). Solar-based controlled system is the most up-to-date kind of solar-oriented cooler that needn't bother with any solar-based boards or battery as it depends on the evaporative rule. A day-to-day existence cycle cost investigation indicated that solar based, and lamp oil coolers cost generally the equivalent yet that solar-powered fridges are liked as they are more solid and environmentally practical.

Figure 2.9: This is not common in small-scale growers and has other disadvantages, such as the high emission capacity and increasing recurrent costs. Diesel power systems can be utilized for certain larger and mid-size off-grid farms.

Source: https://www.esmap.org/empowering-farmers-in-africa-productive-use-levering-solar-

2.9.3 Processing

They should be dried prior to packing and advertising to increase the value of fresh leafy foods. The utilization of low warmth and ventilation to the most awesome aspect the outcomes are adequate for the evaporating of new items containing to 95% water, with a solid stickiness substance of 7 to 8%. Subsequently, limited scope preparing of foods grown from the ground is less motorized than other cultivating practices. The gear utilized is produced using treated steel since organic product acids are produced using mellow steel. For vast drying, it is generally essential to strip, cut or slice formed into uniformly estimated pieces. This is normally performed manually. For larger amounts devices are accessible which can accelerate peeling, cutting, or processing of the goods. The heat-supported drying with a wide range of dryers are used for drying materials (e.g., abaca, copra, coffee, or tea).

Laying things to dry in the sun is an extremely modest drying strategy famous in bright environments. It stays the principal strategy for the overall creation of grapes. In any case, customary techniques for sun drying regularly yield low quality on the grounds that the merchandise are not protected from downpour, dust, and different poisons. Different sun-based drying requires a covered dryer that shields against direct daylight while warming from the sun. A circuitous dryer diminishes warmth and harm, while quickening the drying with a regular wind stream inside.

The warmth helped dryers create a warm wind stream inside the dryer in blustery climate zones or in clammy spots for quick drying. Electrical, propane, wood, or other fuel accessible locally can be warming sources. With low-fueled, moderately modest innovation, adequate drying temperature can be effectively accomplished. The greater part of the warmed air dryer energy is utilized for cooling (Raymond et al., 2019). Moving air energy is only a little part of air warming costs that rely upon the item's underlying moisture substance and last moisture content. If the cycle requires energy-saving advances, for example, air distribution, the fuel use will diminish fundamentally.

Experiences and examples from peanut production in Bolivia: Edible oil is a high-esteem pay yield item. The main oil development items are nuts, for example, palm bit, coconut, shea nut, groundnut, seeds of oil (mustard seed, sunflower, sesame, and so on) and palm oil. Oilseed side-effects incorporate high protein press cake and creature feed supplement content, while coconut and groundnuts are human results. For ranchers, the two of them produce additional benefits. In the biogas digesters, results of agro-mechanical oil plants are utilized to deliver power or interaction warmth or steam.

The pressing of oil (spindle presses), hydraulic presses or oil presses with engine control are also manual technical features. Mechanical control alternatives are very demanded to minimize human drudgery. A method of bringing small-scale oil expelling firms to rural areas in Africa has been made by scientists and technology specialists working with Appropriate Technologies International (ATI).

A little cylinder, which presses a deliberate heap of oilseeds into a metal enclosure is utilized in the hand-held smash press. In 1985, Carl Bielenberg of ATI made the essential plan in Tanzania. In this locale and somewhere else in Africa, unique and improved models are presently generally utilized. The dry extraction technique must be finished by mechanical or water powered presses, yet in addition by consistently running oil expellers (Simoes et al., 2014). Exporters may regularly be fueled by animal or waterpower utilizing petroleum motors or diesel motors or electric engines.

IBG Montforts + Reiners, Moenchengladbach, Federal Republic of Germany is fabricating a little mechanized expeller called KOMET. The substance of the enormous size ought to be cut into 5 mm width parts, for example, copra and palm portions. A 3-kW engine with a nonstop factor gearbox is the standard force source. The machine can likewise be controlled utilizing a dynamo-grasp prepared gas motor (11 kW at 3000 rounds). When all is said in done, countless assembling hardware and apparatus makers exist in Africa, Latin America, and Asia. To redesign traditionally limited scope innovation, in any case, interest in innovative work (R&D) is required. To overhaul traditionally limited scope innovation, nonetheless, interest in innovative work (R&D) is required. For instance, in sub-Saharan Africa simple innovation has prompted an ascent in metropolitan reliance on imported rice and wheat for sorghum, millet, and other nearby grains. Admittance to crude material, inputs, for example, packing materials, apparatus and extra parts and expert HR are additionally basic variables for effective preparing exercises.

2.10 COMMERCIALIZATION

The energy contribution to capacity and treatment plants is recorded previously. Rural power framework venture is not only essential for advertising when all is said and done for foundation improvement. Rural foundation configuration should make it straightforward for smallholders to enter nearby business sectors and lessen transportation and appropriation costs. Manual vehicle advancements comprise chiefly of vehicles, trucks or carriages driven by creatures. The most mainstream method of transport is given by petroleum products or biofuel engines for vehicles or trucks (Shove & Walker, 2014).

Market information is basic for ranchers and brokers to have the option to make educated choices about their yields, their harvests, markets and whether to store them. Market data is basic. New data and correspondence advances (ICT) open freedoms to advertise data administrations to improve data supply through wireless SMS and give more limited data benefits by quickly developing FM (recurrence balance, broadcast innovations) radio broadcasts. Cell phones frequently incredibly lessen the expense of contact and data by limiting distances.

In India, an exploration study found that the extension of the inclusion of cell phones brought about a considerable fall in use costs (4%), while anglers were expanding their benefit by 8%. In the long haul, the Internet will turn into a proficient method to furnish ranchers with information. PV-fueled cell phones and PV satellite telephone booths have a standard force pinnacle of 2 - 3 kW and are outfitted with a 50 WP gadget with 2 lights and a force attachment to charge the batteries for the cell, just like the case with Bangladesh-Grameen Shakti.

Ranchers, merchants, and others in various chains acquire a comprehension of the social, mechanical, monetary, and ecological elements which sway the insurance, quality and estimation of the farm creation and advertising. The inventory of power to structures, including practical sources of energy, for example, sun-based lighting or energy-effective structure materials, would make them a model of present-day energy administrations (Sørensen, 2008). Farm lighting with the PV/battery framework incorporates a force supply in the reach 50-500watts, for instance (for example, security and security of scattered structures). Notwithstanding TV, faxes, power lights, and so forth, PC frameworks need 8-300 Wp for preparing. Internet e-commerce servers are primarily installed into a multifunctional solar system (1 kW).

For practical promotion, it is essential to have a proficient advertising foundation, for example, gathering markets, retail markets, and discount markets. In significant metropolitan territories, terminals discount markets are found and the items, at last, can be shipped off buyers through exchange between wholesalers, retailers, cooks, and so forth as changes in retail responses to metropolitan development, the developing situation of stores and expanded client buying power, discount market qualities have changed altogether. These improvements incorporate responses to the association and the board of traditional discount markets. The country gets together business sectors are situated underway zones and serve essentially as spots where farmers can meet with brokers to sell their items. These business sectors might be infrequent (maybe week after week) or perpetual, regularly occurring around evening time or in the early morning when lighting is required. On the off chance that there is no association with the public framework, utilization of sun-based lights can be another option. PV battery-powered fluorescent lamps range somewhere in the range of 10 and 20 watts for each light.

Figure 2.10: In our culture, farmers have often performed an important role in providing food to the world's people. Even so, you can overlook that it provides not only nutrition and energy, which today, particularly considering sustainable power, is of the greatest priority.

Source: http://www.renewableenergyfocus.com/view/44586/solar-powered-irrigation-a-solution-to-water-management-in-agriculture/

2.11 FUTURE EXPECTATIONS

Even if direct energy were just 1.1% of the energy used in the United States over 2002 in agricultural activities, we could see that the property used during the cultivation of 914 million acres generates considerably more energy than the accumulated amount. Such variations in energy quality mean that improvements or levers may be made to increase efficiency. Firstly, year-on-year growth in yield was achieved by better inputs and techniques in small single-digit numbers. Secondly, urban, and vertical farming, which increases the use of land resources that concentrate crops and free soil. Eventually, there are major variations in power storage between crop and crop, as we saw in soybean and corn. In summary, the social environment in farming will shift dramatically in the coming years as sustainable development is requested.

It is not too necessary to stress the significance of farming in the world of man; it is the sole supplier of food for humans. With the awareness of their value and importance, initiatives should be continually implemented to promote their survival and to address their problems (Silvente et al., 2015). The use of renewable energy sources is one of these steps.

Here are some important functions in maintaining sustainable agriculture for renewable energy:

- Solar energy –Solar energy is limitless and supportive to the environment. Solar- powered energy innovations are relied upon to be basic to a wide scope of warm applications and decentralized energy age and conveyance frameworks soon. Sun based energy innovations incorporate solar oriented photovoltaic frameworks, sunlight-based water warmers, Solar-based cookers and solar-based boards. Solar oriented boards are likewise accessible.
- Wind energy –Growers and producers can produce their own wind power by using small wind turbines ranging from 400 watts to 40 kilowatts or more. Such wind turbines can be built for the needs of a whole farm or for application to save electricity production prices.
- Hydro-energy - The drop in the water turning cutting edges of a turbine produces hydroelectric force, which is associated with a generator changing over hydro electrical energy into power. Hydroelectricity is a spotless homegrown, sustainable power source that provisions zero emanations with ease power. It is the lone sustainable power source that can substitute power age from petroleum products and meet rising energy necessities.
- Biomass - Subsequently it is conceivable to limit various issues identified with biomass the executives and use in homegrown and enterprises, utilizing procedures, for example, scorching and briquetting (pressure of biomaterials for fuel). The deposits of the yield would now be able to be briquetted and utilized for cooking for kindling or other non-renewable energy sources.

Along these lines, it is conceivable to limit various issues identified with biomass the board and use in homegrown and enterprises, utilizing strategies, for example, scorching and briquetting (compression of biomaterials for fuel). The buildups of the yield would now be able to be briquetted and utilized for cooking by firewood or other petroleum products.

2.12 MODERN ENERGY LEVELS OF ACCESS

Admittance to energy in agribusiness is significant for guaranteeing that farming creation will fulfill the expanding need for food. Attributable to the absence of street framework and a power matrix, some little, separated rural networks stay without accesses to present-day energy administrations. This deficiency influences the creation and handling of food as well as the planning of staples, which concentrate a lot of energy utilized on the natural pecking order, especially in most unfortunate nations. In developing nations, conventional solid agriculture such as wood fuel and cattle dung provide most household resources. Women and children are typically responsible for gathering and supplying domestic energy with negative health implications because conventional biomass-induced cooking is related to indoor air contamination and heavy fuel consumption. There is also a significant time to gather wood fuel, from less than an hour to almost 8 hours a day.

Even if power supply lines are established, the supply can be extremely unstable with regular failures and varying power performance This type of plant also employs diesel generators for electricity production or implements renewable power systems (such as small-scale wind or hydro power systems). Power gaps are exceedingly challenging to measure in small farmers and associated rural businesses, as sources and applications of power are so diversified and diversified, distributed across millions of small farmers and populations. However, some leading statistics offer a glimpse of the issue size and in Sub-Saharan Africa indications of this (IIED 2014) can be discovered:

- The majority of the ranch power depends on human exertion (65 percent) or animal force (25 percent), with a minority from motors (10%) — much lower than for other arising zones, where motors establish 50% of farm power.
- In contrast to 39 percent in South Asia and 29 percent in East Asia, only 4 percent of the cropland is irrigated.
- After processing it, an approximate 10-20% is lost at an annual expense of $4 billion, which is equivalent to the number of cereals imported annually.

Chapter

3

Choosing the Right Energy Source for Agriculture

CONTENTS

3.1 INTRODUCTION

In most economies, especially the developing ones, agriculture is considered to be the most dominant sector. As such, it is given much attention by the agents of a country to ensure that it is always present to earn income. Considering the gross domestic product of most developing countries, the sector that contributes the majority share of income is agriculture. Ideally, agriculture continues to be the source of income, livelihood and even employment in the stated countries. Furthermore, the sector is considered to be dominant since it is depended upon by 50-90% of the total population in developing countries. With the above facts, it is with no doubt that agriculture in developing countries plays a vital role in the economic development of a country (Silvente et al., 2012). This statement is with no doubt truth because, with agriculture, there was and still is a significant contribution to the prosperity of most of the already developed countries. In some of the big countries such as Europe, the USA and Japan, it is clear that agriculture was introduced before the industrial revolution and it helped to a great extent in the process of industrialization. As such, many developed countries have learnt that the only way to attain advancement is by understanding that agriculture and industrialization are complementary sectors. One sector cannot lead to the success of a country without the other. As such, with the industrial revolution, agriculture is still at the hearts of many members of society.

Figure 3.1: Agriculture plays a vital role in developing countries to improve their economies.

Source: CIO East Africa

With increased agricultural output and also productivity, the results are very much comforting as it contributes to the general economic development of a country. As such, it is considered to be only rational for developing countries to put much emphasis on the development of the agricultural sector. There are various ways in which agriculture makes its contribution to the economic development of a country. One of the ways is that, with agriculture, the individuals living in the rural sectors who depend solely on farming of agricultural production, are able to demand goods that are not produced in the agricultural sector using the money that they are able to earn from the selling of their produce. Agriculture plays a vital role in providing food and raw materials for other sectors that are non-agricultural. With agriculture, the country can be able to earn foreign exchange by selling their products as exports to other countries. Individuals who are practicing farming of agricultural produce also have the ability to have investments in the form of savings and taxes especially when investing in the non-agricultural sector (Utama et al., 2014). The individuals that are most vulnerable in the society in terms of not having received any form of education and the ones with unskilled labor tend to depend majorly on agriculture as their source of income. This shows that agriculture is a source of employment for such individuals.

There are also ways in which agriculture plays a vital role in the economic development of a country. One role played by Agriculture is that it contributes to national income. From what is learned from the history of most of the already developed countries, the takeaway is that most of them prospered in terms of economic development because of the fact that they took part in fostering the agricultural sector. The second role that agriculture plays in the economic development of a country is that it acts as a major source of the food supply. Food is very important in the everyday life of an individual whether in developing countries or advanced ones. Agriculture on the other hand acts as a basic source of the food supply. With the increasing population sizes all over the world, the dependence on agriculture is wide. In the cases where agriculture fails to meet the rising demand for food, the consequences are harsh especially to the growth rate of an economy. Agriculture is also considered to be a pre-requisite for raw materials. Most finished products need agricultural products especially, in the agro-based industries. Ideally, the shortage of agricultural produce is seen to greatly impact the industrial sector. Some of the industries that majorly depend on agricultural produce include rice shellers, oil and dal mills, flour mills, milk, bread, meat products, wineries, sugar factories, textile mills, jute mills among many other industries. Another way in which agriculture plays a role in the economic development of a country is by the provision of surplus. With the availability of surplus, the country has the ability to sell the surplus as exports. This increases the exports that are sold from a particular country. With exports, the country will have the ability to earn foreign exchange which can be very important when there exists a situation whereby a country is in a dire need to access some of the basic and essential capital goods.

Most developed countries are normally faced with the burden of having a shortage of foreign capital. Agriculture tends to require less capital in its development and thus it minimizes the problem of foreign capital. With agriculture, there is also the development of infrastructure. This is so because, the development of agriculture requires that there is the development of roads, railways and even postal services. With the development of infrastructure, there will be an increased demand for industrial products. Most countries that depend largely on agriculture and also overpopulated tend to experience greater inequality in terms of income and as such, there is a need of putting emphasis in the development of the agricultural sector (Uchiyama, 2002). With the growth of the agricultural sector, there is an increased demand for both products from the agricultural sector and from the non-agricultural sector. In most developing or rather underdeveloped economies, the majority of the population depends largely on agriculture in order to allow them to afford the consumption of products from the non-agricultural sector. Another important role played by the agricultural sector is that it tends to be very helpful in phasing out the economic depression. When there is a depression in the country, industrial production can be reduced and in other instances can be stopped but the agricultural sector can be left standing. With these situations, the agricultural sector will be able to create effective demand at such adverse conditions in the country.

Figure 3.2: The development of agriculture leads to the development of infrastructure.

Source: Raconteur

With the above contributions of agriculture to the economy, the sector has been put in the spotlight and is currently being modernized. Modern agriculture itself requires modern energy. Agriculture and energy are closely linked. This is so because modern energy is helpful in the development of the agricultural sector. The introduction of energy in modern and sustainable agriculture, its production and processing system will help in the movement of agriculture from subsistence farming to a place where agriculture ensures food security. With this, there will be added value to the rural areas and the expansion into new markets preferably the international markets will be realized. Some of the known techniques that provide energy, which substantially support the agricultural sector and the various production processes include renewable energy technologies and hybrid systems. These energy systems support agriculture by the provision of postharvest systems such as those used in the cooling of agricultural produce, ensuring that there are proper irrigation systems through the provision of modern-day pumps and also the processing systems such as the dryers, millers and pressers.

When using energy, it is also considered to be beneficial to society since most of the production processes require human labor for them to function. Animal labor is also not excluded; it is very beneficial to the production process. For the entire process to be successful, fuels are required to perform some of the known major functions such as pumping, mechanization and other activities. All the above processes are also indirectly helpful in the production of agricultural products as they are used in the production of agrochemicals and fertilizers which have proven to be very useful in farming.

The chain of production has a certain chain that involves a number of steps. The first step being production, the second being postharvest and storage, the third being processing and the fourth being the commercialization of the produced products or rather the finished products (van Ruijven et al., 2019). All the above steps require energy in order for them to function effectively. However, the production process is the one that requires the most energy. In modern society, modern technologies have been invented in order to be helpful in the tilling of land and also for planting.

3.2 HOW AGRICULTURE UTILIZES ENERGY

In our everyday lives, researchers and scientist have been trying to come up with measures that can be used to reduce the emission of toxic gases to the environment. Such gases contribute to global warming and also the pollution of air that is considered to be very crucial to the human population.

The very first suggestion provided by most of the scientists is buying a car that has the ability to compost, recycle energy or one with better gas mileage or buying appliances that are considered to be energy efficient. All of the stated methods are considered to be very good in the quest of reducing the consumption of energy in the world.

In most developed countries such as the USA has the agricultural sector contributing a significant percentage to the emission of greenhouse gases. In the quest of solving the energy crisis in the world, one should start from the agricultural sector. Farm equipment to begin with consumes energy. In each stage of farming, there is the use of machinery and equipment which burns fossil fuels. The increased greenhouse gases in the air are added when these fuels are burnt. Some of the gases produced include methane, nitrous oxide and majorly carbon dioxide, which leads to the increment of the earth's temperature. When thinking of equipment that is used in the firm, tractors are what comes to people's minds (Wang et al., 2011). However, this is not the only machine that is used in the farm. It is just that others are not given much attention. Take for example the case of slaughtering an animal, water is not considered to be a variable but as a matter of fact, it is considered to be a very important factor.

Figure 3.3: The burning of fossil fuels leads to the pollution of the environment.

Source: INSH

When slaughtering even a single animal, a large amount of water is used. What should be noted is that the water does not just appear from thin air, it needs to be pumped in out of the slaughterhouse in order to make work easier. Otherwise, one would be required to hire manpower to be carrying the water in and out of the slaughterhouse which can be very expensive and quite inefficient. The pumps that are used to bring water in out run on a form of electricity be it grid, wind, solar among others. Electricity on the other hand runs on fossil fuels. Considering the fact that billions of animals are being slaughtered each year, the energy usage is considered to be truly outstanding. The demand for meat all over the world is quite high and it may prove to be quite difficult for the farmers to control the demand. However, there are ways in which the farmers can adapt to reduce the dependence on fossil fuels in farming. This is especially efficient to farmers whose majority of their farms are reserved for no-till. To begin with, farmers can decide not to cultivate soil when they are planting seeds which can be very efficient in reducing soil erosion as well as reducing the amount of diesel used since there will be limited use of machinery in the cultivation process.

It is not only in cultivation that there is the consumption of energy. The process of processing packaging and the transportation of agricultural produce is seen to be the major users of energy. The production of food be it cultivating crops or the slaughtering of animals is considered to be a small part of the equation.

The entire processes that food takes before it gets to a consumer's plate are variables to be considered when looking at the dependence of agriculture on fossil fuels. To begin with, the fuel that is used in the transportation of either agricultural raw materials or even agricultural food products is very enormous, especially when dealing with international marketing. Even if countries decide to reduce this by consuming products that are locally produced, there would still be a sector of agriculture that will act as a way of wasting energy (Waibel et al., 2019).

The area would be packaging. Just as seen above that a lot of energy is required to pump water that is required in a slaughterhouse, the energy that is required in the packaging of food products in order to make them fit for consumption whether internationally, locally or even across the street is quite high. To reduce the carbon footprint, the farmers can shift to the use of preconditioned bags that are useful in the transportation of agricultural produce.

Figure 3.4: Transportation of agricultural produce also consumes energy.

Source: ESSPL

Most of the products that are purchased from the supermarkets are normally imported or shipped in locally and in most cases, farmers would use plastic bags. Initially, the farmers would burn these bags or even place them in trash cans. In the current society, plastic bags are produced in limited quantities and the use of such bags is even prohibited in some countries. With such measures put in place, the farmers are forced to reuse the bags which are generally environmentally efficient, reduces the amount of waste that fills the lands each year and also reduces the workloads to factories that create such products. As such, the factories of plastic bags have been forced to venture into the production of other products.

There are also various ways in which agriculture indirectly consumes energy. This is mostly in the production of non-agricultural produce such as fertilizer production. For plants to grow, there is a need for photosynthesis to take place that utilizes nitrogen. This means that the soil needs to have nitrogen in order for the growth process to take place. Soil that is rich in nitrogen yields better harvest which is exactly what is needed by most if not all farmers. The easiest and most efficient way recognized by farmers to increase the amount of nitrogen in the soil is by application of fertilizer. In order to have large amounts of food in the world, there is a need for using large amounts of fertilizers. The use of synthetic fertilizers by farmers all over the world has become the option of the day in order to make their soils nutrient-rich in order to yield more crops in a limited size of land. Approximately, millions of tons of fertilizer are used in the world every year. However, everyone knows that such inputs do not appear just from thin air. They need to be produced and there is a cost that needs to be paid. To produce synthetic fertilizers, manufacturers use non-renewable sources mainly fossil fuels. It may deem unproduceable with the absence of fossil fuels. The production entirely depends on the burning of fossil fuels. As seen before, the burning of fossil fuels leads to an increase in the earth's temperature that shows that they contribute a larger percentage of global warming emissions. Ideally, there is a high demand for synthetic fertilizers, which is responsible for damaging the environment (Weiss et al., 2010).

In such circumstances, the farmers should be schooled on the various ways of achieving high yields without depending on synthetic fertilizer. In order to increase the amounts of nitrogen in the soil, farmers can engage themselves in the planting of nitrogen-fixing plants at the points where they are harvesting their other crops. Some of the plants that can be planted include the alfalfa and the soybeans after harvesting the major crops since nitrogen-fixing bacteria is found in the roots of these crops. Once the plants die, the nitrogen is released to the soil that can then be freely utilized by the plants that will be planted. Simply stated, this way, the farmers will be able to introduce nitrogen in the soil without burning fossil fuels.

Figure 3.5: Most farmers use synthetic fertilizer to improve the amount of nitrogen in the soil.

Source: TopTillers.com

Ideally, it is with no doubt that agriculture consumes a lot of energy that is believed by most members of society. With the changing environmental conditions, the majority of government stakeholders are placing more restrictions on the emission of carbons which is acting as a restriction to the activities that consume a lot of energy. Farmers on the other and should look at government concerns as a breakthrough and also do their part in reducing carbon emissions. By doing this, the majority of farmers will minimize their dependence on fossil fuels which to a greater extent will save more of their money. If members of the public become more considerate in how they produce, how they consume food and the transportation used in moving food from one point to another, the reduction of greenhouse gasses and emissions will come naturally. In instances where there is the need to use energy sources or rather power, farmers and even industrialists should be sure to choose sources that are environmentally friendly and also people-friendly. The source should be one that does not highly emit gases and also cost-effective. Below are some of the energy sources that can be used in the various stages in agriculture whereby one can be able to see the methods that are both economically efficient and also environmentally friendly when used.

3.3 GRID ELECTRICITY

In understanding how energy works there is a need to understand what is meant by gird electricity. To begin with, a grid is recognized as a network of transmission lines that is usually used in the distribution of power. Power has proven to be very beneficial in the majority of households in the world and for safety transmission, there is a need for grid lines. Grid electricity is whereby power is transmitted by an interconnected network that is solely made for the delivery of electricity from the producers to the consumers of the same. Electric power is produced by generating stations and the voltage is stepped up for transmission or down by distribution by the electrical substations. The distribution lines vary in size depending on the consumer of the power transmitted by the distribution lines. In order for all this to be possible, most of the power stations that are connected to the electrical grid are mostly located near an energy source or rather a resource including a source of fuel. In other circumstances, they take advantage of the renewable energy resources which are preferably further away from areas that are densely populated. In cases where there is a need to transmit power for longer distances, bulk power transmission networks are considered to be quite useful and can be quite reliable even when crossing international borders (York, 2016). In such circumstances, electric power is normally stepped up to a higher voltage in order to meet the high demands for power all over the world. The stepped up power normally passes a substation which then steps down the power to a transmission level in order to distribute to various households.

Figure 3.6: Power is transmitted by an interconnected network is known as grid electricity

Source: Electronic Design

This form of energy is majorly used to provide power to run industries, to light up households and run some of the small equipment found in the households such as the fridges and ovens. Since electric grids are prone to attacks or even malicious intrusion, there is always a need for electric grid security. Apart from the above-stated functions of grid electricity, they can be very useful in agricultural development. Electricity as an agricultural input can help in improving productivity and better yet improve the incomes of the agricultural workforce.

In rural areas, electric grids are considered to be a key infrastructure that contributes to the living standards of farmers. It is claimed that grid electricity has a substantial impact on agricultural production and is very useful in rural prosperity. Since the year 1998, China has been able to see the benefits that relate to the use of grid electricity in agricultural production. As such, there has been the implementation of the policy and furthermore, the electrical system has been further modified in order to ensure that there is better management of the power system. A policy has been established to ensure the success of the same and it follows that in the country every individual is equally charged for the equal net. With the above measures, there has been an improvement of the rural power system or rather the grid structure and its reliability in terms of supply has been notably enhanced, there has been a reduction of the electricity prices in the rural areas which has substantially created suitable conditions for the social and economic development in the rural areas. The power transmitted by the electric grids can be very useful in the pumping of water from the well and also can be used in irrigation. If the electricity is stable, it can be quite useful in the farming of products that need 24/7 care. Maintaining hybrid animals such as the hybrid chicken can be very beneficial.

In the current society, most developed countries are efficiently using grid electricity to safely transport the produce from one point to another without depending on hired labor. In this case, the vehicle is remotely controlled that ensures safety. Also, there is on time delivery of the same. This method is quite efficient because there will be no use of vehicles using diesel that emits gases that are harmful to the environment. Most of the equipment that are used in the farm use diesel which is a fossil fuel that emits gases that are harmful to the environment (Züttel et al., 2010). Many countries have made grid electricity available to most farms that have reduced the use of generators that use fossil fuels. Grid electricity is also quite efficient and cheaper to use compared to the use of generators. Another method in which agriculture acquires energy is the use of fuels for combustion engines.

3.4 FUELS FOR COMBUSTION ENGINES

Combustion engines require fuels for them to function. Combustion engines allow for the drivability and also the durability of the majority of locomotive machines including cars. For these engines to run, they utilize fuels such as diesel or even gasoline. Consequently, they can also utilize alternative fuels such as renewable ones including propane, biodiesel, natural gas and even gasoline. In order to contribute to fuel economy, they can be combined with plug-in hybrid electric systems to extend the range of hybrid electric vehicles or even with hybrid electric power trains. These combustion engines are also found in farm machinery such as tractors and the combined harvesters. Some of them have larger combustion engines. These combustion engines are very important in running the vehicle. To understand this, one should understand how they work. Combustion is basically the process involved in burning. It is considered to be a chemical process whereby there is a mixture of and fuel in order to release energy. In a combustion engine especially internal combustion engines, the burning of fuels takes place within the engine (Armaroli & Balzani, 2007). The energy that is produced is then basically converted to work which is the main reason as why a harvester has the ability to move in the farm while harvesting and the tractor to till the land while moving. A fixed cylinder and a moving piston that makes up the engine are what make the entire process take place. Combustion gases first expand which in turn pushes the piston which then rotates the crankshaft. Following a system of gears in the locomotive machine such as the tractor, the motion drives the wheels and the other working systems such as the plow.

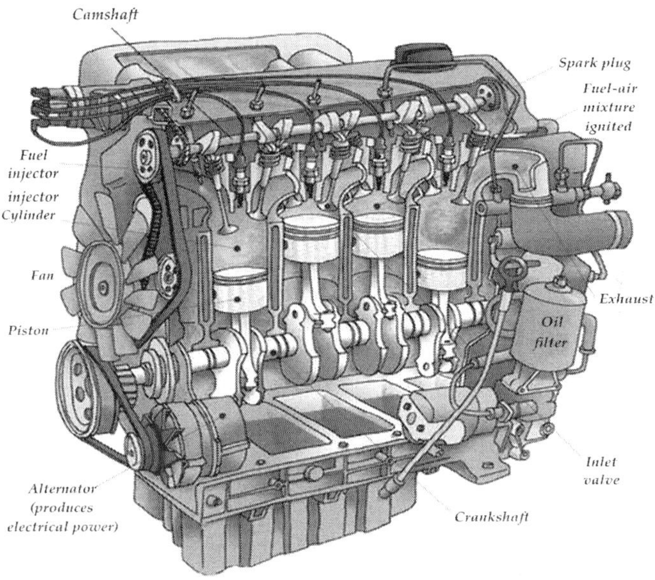

Figure 3.7: Combustion engines are useful in the drivability and durability of locomotive machines.

Source: Q-files

Over the past years, researchers have been working hard to try to reduce emissions that are experienced by the use of machines that use combustion engines such as nitrogen oxide. The aim of this is to protect the environment while also maintaining or rather increasing the fuel economy. Mechanization of agriculture basically utilizes diesel internal combustion engines. These engines require energy that is stored in liquid form. For there to be an improvement of the fuel economy, then there is a need for partial substitution of diesel fuel that is of mineral origin to the use of that from plant and animal origin commonly known as biodiesel.

The majority of economies are driving towards the use of renewable fuels. This is due to the growing concern over the environment with regards to pollution and also the growing prices of oil. Currently, researchers are using non-renewable fuels together with renewable fuels. This is considered to be useful especially as they are hoping for a transmission from the use of non-renewable sources to renewable sources. Agricultural production is quite high in most parts of the world and the use of machinery such as tractors, harvesters, self-propelled sprinklers has skyrocketed meaning that there is an increased use of fuels that is unfriendly to the environment and also the fuel economy. The advice is to slowly shift the attention to the use of biodiesel which can be easily procured from animals or plants and has very minimal implications to the environment when it are used. Fuels for combustion engines are quite important as they run most of the heavy machinery used on the farm. Without them, then it means that there will be a need to hire human labor which has proven to not be reliable especially in the current society of democracy where people cannot be trusted. Another source of energy that can be utilized in agricultural production is the use of solar energy (PV).

3.5 SOLAR ENERGY (PV)

Solar energy (PV) is what is termed as photovoltaic solar energy. It is considered to be a clean and renewable source of energy that utilizes solar radiation to efficiently produce energy. It is named based on the photoelectric effect in which electric current is generated in the process by which certain materials are able to absorb protons which are basically light particles that release electrons. For this purpose, a semiconductor device known as the photovoltaic cell is used. The use of solar energy can be quite limiting especially in sectors such as agriculture since some of the equipment used is not made to be run by solar energy. The use of solar energy also is limited by a number of factors. One of the being sunlight availability. The production of solar energy depends solely on sunlight. Weather structures are one thing that cannot be controlled by human people (Anishin et al., 2013). There are instances in which there is no sunshine and in such circumstances, you better be equipped by other methods of acquiring energy. As an agriculturalist, one should be wise enough not to depend on solar energy since at any time the weather might change.

Figure 3.8: Solar energy PV is a clean and renewable source of energy.

Source: Student Energy

Another limiting factor of solar panels is the installation area. For homeowners, the installation area may prove not to be challenging but when dealing with large companies, there is a need for a large number of solar panels and therefore require a large area for installing the panels. Another limiting factor is reliability. As is known, solar energy relies entirely upon on the sun and in the night, they can't be utilized. As such, a company would be required to connect to an alternative power source such as the electrical grid so as to have an energy source to use during the night. Clouds and storms can also restrict the amount of energy that can be produced by the solar panels as they block sunlight from reaching the solar panels. Even though the use of solar panels can be highly recommended because of their limited negative effects on the environment, there are some impacts on the environments that are recognizable such as land and water use habitat loss and the fact that hazardous materials are used in its manufacture. It is also considered to be a system linked with expensive energy storage especially when producing power at an industrial scale. In the current state, people resort to the use of batteries to store solar energy as the use of electrical energy has been considered to be quite expensive. A company known as Tesla has even gone to an extent of creating the Power wall battery that stores solar energy to be used for future use.

Even though there are disadvantages to the use of solar energy, the advantages have been realized, especially when it comes to agricultural development. One is that they are cheaper and safer compared to the use of fossil fuels. It is with no doubt that the installation might be quite expensive with solar energy one can be able to avoid grueling costs when it comes to maintaining one's farm. During the drought, it is considered to be quite sensible as it requires minimal amounts of water delivery. While solar energy especially the photovoltaic panels produce smaller yields, during the drought it might be the best solution possible. Another thing is that solar energy is not a permanent solution. This means that if a farmer has installed energy panels, they also have an option of reverting to the use of electricity. During the drought, farmers can easily utilize solar of which if they decide that it is good for them they can easily revert back (Ahmad & Zhang, 2020). With more companies aiming at using energy sources that are friendly to the environment, the use of solar panels has gained popularity in the business world. With widespread acclaim, more advancements are made on it in order to meet the need of the users and since the majority of the agricultural producers and processors use solar energy.

Figure 3.9: California farmers are planting solar panels with the decline in water.

Source: Daily Republic

Over the years, the storage of solar energy is becoming more and cheaper. With this insight farmers, especially those from California will be very happy since the area is prone to natural disasters. The preservation of solar energy having in mind that there may be a crisis in the future can be very helpful and is considered to be an advantage of solar energy. Another advantage of solar panels is that it makes the soil to be a better habitat for both plants and animals. Most areas in which the solar panels are installed are surrounded by grasslands and it does not seem to be affecting the surrounding vegetation. It is also not hard to maintain the panels as they can easily be maintained by hedging which poses no threat to the surrounding vegetation. As such, animals are able to get more grazing land as it creates a pasture-like setting. The assurance is that the animals that are chosen to graze in these areas cannot be harmed by either the underground or even the above-ground panels. One of the most important advantages of solar panels that have made the majority of individuals switch to the use of solar energy is the fact that the use of solar panels is considered as the most efficient way to halt global warming. The utilization of renewable energy has been recognized as a productive method to reduce carbon footprint. It is widely known that the way of acquiring renewable and large amounts of clean energy is by the use of solar panels.

The use of solar panels has so far been considered an efficient way of acquiring energy. One thing is that its effects on the environment is quite limited and is also economically efficient as it is cheaper compared to other energy sources. As such, manufactures of the same should try to come up with better materials and ways to make solar panels as those are the major limiting factors to the use of solar energy. Another source of energy that can be used in agriculture is wind energy (Basore et al., 2018).

3.6 WIND ENERGY

This is one source of energy that is not usually considered by many producers and processors as they are considered to be highly unreliable. It involves the use of wind which provides mechanical power through what is known as wind turbines which then turns the electric generators to produce electrical power. It is a renewable source of energy and thus can be considered o improve the economy of fossil fuel while in the process protecting the environment. It is considered to have a much smaller impact on the environment compared to the burning of fossil fuels. The wind is said to be an intermittent energy source. This is because of the fact that they cannot be dispatched whenever one demands of it. The energy power that is provided by wind is said to be consistent year after year though it provides power that is greatly varied. As such, most of the scientists claim that it gives variable results. Just like solar power, wind power can also be stored which tends to reduce demand, especially in times when wind production is quite low. Upgrading of a grid in a certain region can be done when the wind energy distribution in the region starts to increase. In the world, about 5% of electrical generation is supplied by wind. To capture wind energy, there are wind farms. Wind farms that are located onshore tend to have significant visual impacts on the landscape. Unlike power stations, wind farms need to cover large sizes of land and also need to be built away from densely populated areas. This is because, for wind energy to be more efficient, there should not be any form of blockage to allow for the wind to be captured in order to drive the turbines.

Figure 3.10: Wind energy drives wind turbines which turn electric generators to produce electric power.

Source: National Geographic Society

Wind power has no fuel cost and but is considered to be capital intensive. As such it is considered to be quite stable in terms of pricing compared to the use of fossil fuels. For centuries, the grinding of grain into flour and the pumping of water from the wells utilized wind power. In the current society, there has been an increased support for the use of renewable energy. Most farmers are using their farms as wind farms while also planting crops in the same land. However, the crops should be shorter meaning they should not reach the wind turbines. Harvesting of wind energy from the farms has allowed farmers to also thrive from them furthermore, they are considered to be money savers as in the current society agribusiness can be powered by wind energy. Another source of energy that can be used in agriculture is the use of hybrid systems.

3.7 HYBRID SYSTEMS

The transformation from traditional to modern agriculture calls for modern energy as the engine for many developing countries where agriculture is the domineering sector in terms of economic development. Increasing productivity by the modernization of agricultural production systems is an essential driver in enhancing food security status and improved livelihoods. Additionally, it propels the production from subsistence, mixed farming to commercial production. In most cases, renewable energy technologies and hybrid systems are used as sources of energy services in production activities. These hybrid systems are essentially defined as involving several types of energy generation equipment such as electrical energy generators, electrical energy storage systems and renewable energy sources. These are still considered as trends or the emerging technologies in agriculture for their evolution are expected much more in the future. The implications of their continued evolution will be wider applicability, more designs, wide functionality varieties, compatibility between two or more components, simplicity in controlling, monitoring and diagnosing them and also lowered costs in operation (Asif & Muneer, 2007). The hybrid renewable energy systems (HRES) is a combination of at least two or more renewable energy sources or in other instances, a combination of one renewable and one conventional source. HRES can be connected or not to the grid. With regard to standalone applications, choosing the optimal hybrid energy system is a challenging process mainly due to difficulties in determining the best combination of sources, reduce initial capital investment, reliability of power supply, system components, etc.

Multi-Criteria Decision Making (MCDM) is a mathematical decision approach applied in order to select the best choices out of a given lot alternative in both the engineering and managerial levels of analysis. The choices made out from these mathematical models may however not always indicate the best option. The multi-criteria decision-making approach can be divided into two prominent categories that are outranking methods and compensatory methods. Compensatory methods involve choices that are more preferred owing to their positive attributes over negative ones, thus indicating a nominal tradeoff between the alternatives. Compensatory methods include Analytical Hierarchical Process (AHP) and Fuzzy Logic Decision Making (FLDM). Major applications of compensatory methods in the agricultural sector are in the water resources, like an evaluation of rural water supply, resilience in water scarcity and evaluation of desalination plants. On the other hand, outranking methods follow a sequence of logical decision making steps based on weights assigned to an individual criterion. Commonly-used outranking methods are the weighted sum and product method, the Technique for Order or Preference by Similarity to Ideal Solution (TOPSIS), the Complex Proportional Assessment (COPRAS) method, the Preference Ranking. Organization Method for Enrichment Evaluation (PROMETHEE), and Elimination Et Choix Traduisant la Realite Method (ELECTREE). However, the PROMETHEE method suffers from the problem of rank reversal, which may mislead the decision-maker to choose the best alternatives when compared with other methods.

In the field of renewable energy, the MCDM methods have been successfully applied to energy planning, transportation energy management, utility planning and building energy management. The methods used follows sequential steps involved along with the area of application, strength and weakness respectively from the literature .the steps include: defining a system with the objectives to be met, finding all the criteria affecting the system based on objectives, seeking alternative systems for feeding need of objectives, setting priority or weights to the alternatives, selecting the most appropriate Multi-Criteria Decision Making approach for the use and finding and presenting an optimal alternative for evaluation.

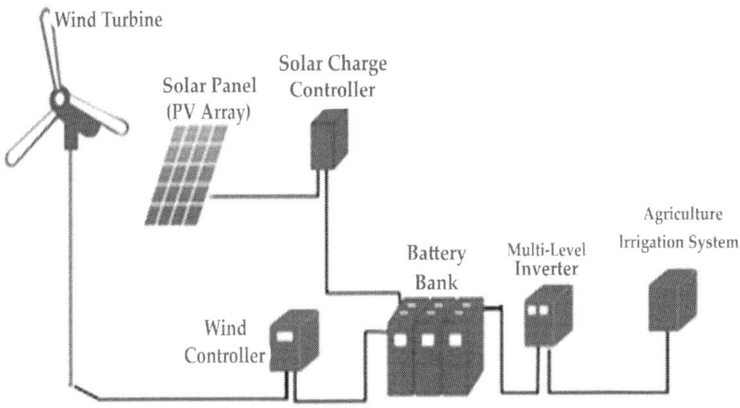

Figure 3.11: Hybrid renewable energy system for agricultural irrigation.

Source: IOSR Journal

The use of a muti-criterion approach in the selection of HRES is currently being adopted because the engagement of a single criterion in the selection of the most suitable hybrid renewable energy system (HRES) has proved and also reported to be inadequate in terms of reliability and sustainability in the agricultural sectors. This follows from the most considered aspects in HRES literature being the technical and economic implications after which the emissions analysis follows. Although reliability, energy efficiency and the use of multiple criteria in system ranking have received low research focus, it has also been reported that many HRES were not able to adequately address the challenges faced by energy needs in the agricultural production in the past because the choice of these HRES alternatives was solely based on a single criterion that is either the technical aspect or the economic aspect.

As such, currently, the selection and ranking of HRES alternatives based on multiple criteria have been preferably considered in production applications. The type of criteria that relate to socio-cultural practices, environmental requirements, enabling policies and so on, are also very important in the selection and ranking of suitable HRES (Batterberry et al., 2010). This ensures that the agricultural operator chooses an optimal strategy to maximize profits. From the technician –economic point of view the "best renewable hybrid system", is one that guarantees the power requirements of the reverse osmosis Autonomous Desalination Systems (ADS), with a daily production capacity of up to 50 m^3 of desalinated water in isolated systems with a scenario of relative high wind and solar radiation profile especially for water scarcity areas.

Electrical energy generators hybrid systems in agriculture are used mainly as backup energy in ensuring continuous production of healthy crops and live stocks in agricultural enterprises.

3.8 BIOMASS ENERGY

Biomass energy is energy generated mainly by living or recently dead organisms and any by-products of those organisms, plant or animal. The term is generally understood to exclude coal, oil, and other fossilized remnants of organisms, as well as soils. In the context of biomass energy, the term refers to those crops, residues, and other biological materials that can be used as a substitute for fossil fuels in the production of energy and other products. From time immemorial biomass energy has been used since the first made of wood fires for cooking and for keeping warm to today where biomass is used as fuel, electric generators and other machinery.

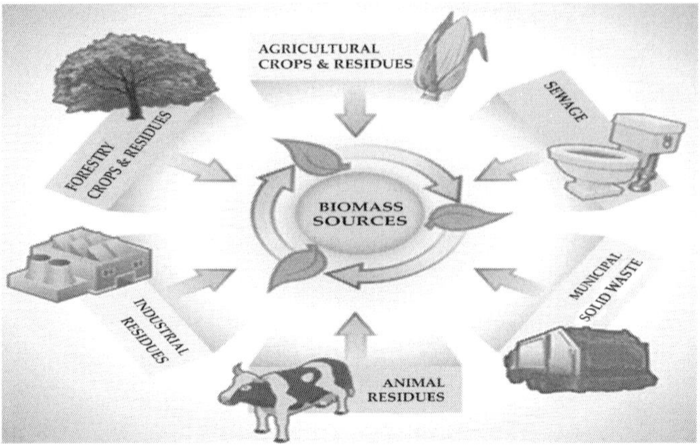

Figure 3.12: Bio mass energy.

Source: Bio Energy Consult

The agricultural biomass which is the matter derived from biological organisms such as corn, straw, plants, animal waste, offal and perennial grasses, as with other types of biomass, agricultural biomass can be transformed into energy and other value-added products, such as bioplastics, medicine, biochemicals, etc. This ensures the most optimal use of the biomass as waste and pollution is reduced. Moreover, utilizing agricultural residues and by-products can reinvigorate rural economies and secure their energy independence. As the global leader in utilizing agricultural residues, Denmark is leading from the front in pioneering the use of agricultural biomass to create energy and high-value products. The production of agricultural biomass and its exploitation for energy purposes can contribute to alleviating several problems, such as the dependence on import of energy products, the production of food surpluses, the pollution provoked by the use of fossil fuels, the abandonment of land by farmers and the connected urbanization (Armaroli & Balzani, 2007). Biomass is not at the moment competitive with mineral oil, but, taking into account also indirect costs and giving value to the aforementioned advantages, public authorities at the national and international levels can spur its production and use by incentives of different nature. However, from a long-term perspective, technological innovation can improve the cost-effectiveness of biomass production. For this purpose, a new cycle of plant domestication is needed, in order to develop new plant genera, species or varieties yielding quantities of biomass higher than plants now cultivated for food production.

Wood is generally the largest biomass energy resource today used by many sectors other sources entailing food crops, grassy and woody plants, residues from agriculture or forestry, oil-rich algae, and the organic component of municipal and industrial wastes. Even the fumes from landfills that contain methane as the main component in natural gas can be used as a biomass energy source. Biomass can be used for fuels, power production, and products that would otherwise be made from fossil fuels. In the agricultural sector, this energy plays an important role in the agricultural and food value chains, especially with regard to the processing of food. Yet, the agricultural and food sector also has a large impact on the emission of greenhouse gases. Due to an ever-growing population, and the increasing scarcity of fossil fuels, the world will need more food that is produced with less energy, or alternative non-fossil-fuel-based energy sources, such as renewable energy technologies. Like wind, solar, and other renewable energy sources, bioenergy can make a positive impact on our atmosphere by lessening our dependence on climate change-inducing fossil fuels.

Biomass or fuel energy comes in liquid, gaseous and solid states and can be used for heating, cooking, processing, cooling, electricity production, and as transport fuels. Most common in the agriculture and energy link, is the generation of electricity from agricultural residues, such as from crops for example the staw and the husk, from husbandry practices such as from the manure residues and from other organic material from excess production or insufficient market, for example, the grass silage. On the consumption side, bio energy can be used for various agricultural processing activities, such as the cooling of agricultural products. When discussing the role of bioenergy in agriculture, cogeneration has to been reported as a relevant technology for the on-site generation of heat cooling thus preventing food losses and for providing electricity for processing purposes which is important in increasing the value-added component, thereby increasing incomes in especially the rural areas and hence optimizing generation efficiency. Besides, only a small portion of the harvested/collected and processed agricultural products are actually consumed by humans and other animals because agricultural residues fluctuate between about 10 and 91% depending on the type of product (Anishin et al., 2013). Therefore, the so-called second-generation biofuels, using agricultural residues, have become a viable option for overcoming energy poverty and granting access to energy in most developing countries. It is estimated worldwide that, about 1.3 billion people continue to live without access to electricity. This is equivalent to 17% of the global population and 23% of those living in developing countries. Nearly 98% of those without access to electricity live in sub-Saharan Africa and developing Asia. However, in rural areas access to energy is crucial for increasing agricultural production, limiting food loss and for ensuring food security. Below are listed some examples of the use of bioenergy and biomass from agricultural residues for the production of energy.

Consequently, in choosing the best biomass for agricultural storage is necessary to accommodate the seasonality of production of the crops so as to ensure regularity in the supply of the biomass utilization plant. The type of storage system will depend specifically on the properties of the biomass, especially the moisture content available. For example, for the high moisture biomass intended to be used wet, such as in fermentation and anaerobic digestion systems, wet storage systems can be used, with storage times closely controlled to avoid excessive degradation of feedstock. The storage systems typically used with dry agricultural residues should be protected against spontaneous combustion and excess decomposition, and the optimal storage moisture depends on the category of storage system employed. Moisture content limits must be observed at all times in generally avoiding spontaneous combustion and the emission of regulated compounds. The cost of storage is an important component in deciding on the best storage strategy for the overall feasibility of the biomass enterprise. In some cases, the storage system can be on the same site as the source of the feedstock. In others, necessary volumes can only be achieved by combining the feedstock from a number of relatively close sources. Typically, delivery within about 50 miles is considered economic on average, but longer-range transport is sometimes acceptable, especially when disposal fees can be down-scaled (Ahmad & Zhang, 2020).

The choice of biomass storage system used at the production site, intermediate site, or plant can adversely affect the cost and the quality of the fuel. The most expensive storage systems, no doubt, are the most efficient in terms of maintaining high fuel quality. Typical storage systems, ranked from highest cost to lowest cost, include: enclosed structure with crushed rock floor, open structure with crushed rock floor, reusable tarp on crushed rock, outside unprotected on crushed rock, outside unprotected on ground and subterranean. The storage of biomass is often necessary due to its seasonal production versus the need to produce energy all year round. Therefore, providing a constant and regular supply of fuel for the plant requires either storage or multi-feedstock to be used, both of which tend to add cost to the system. Reducing the cost of handling and stable storage of biomass feedstock are both critical to developing a sustainable infrastructure capable of supplying large quantities of biomass to biomass processing plants. Storage and handling of biomass fuels are expensive and increases with capacity. The most suitable type of fuel store for solid biomass fuel depends on the space available and the physical characteristics of the biomass fuel.

In conclusion, most of the energy sources used to provide power are considered to be very efficient in agricultural production. Those that utilize renewable sources of energy are however considered to be efficient and safe especially when considering their impacts on the environment. Global warming is a to be a topic that is mentioned by the majority of individuals ranging from the major stakeholders found in the government to the minor members of the society who are the rural poor. The use of energy sources such as solar energy, wind energy, hybrid energy and biomass energy. It is just that fossil fuels are disadvantageous in the environment and are also quite expensive for the majority of the agricultural workers may not be able to afford it (Basore et al., 2018). In their own way, each source of energy can be considered to be the best. Considering the various steps in agricultural production, there are some steps that cannot be achieved without the use of a certain source of energy. Take for example tractors and combined harvesters. These machines run by the use of a combustion engine which produces energy by the use of fuel. In such cases, the only thing for scientists and researchers to do is to adopt the use of renewable energy sources such as the use of biodiesel, hydrogen and natural gas instead of using diesel. All the energy sources mentioned above are very useful and each of them plays a vital role. It is just that the users need to be careful about the fuels they use when applying the mentioned sources.

Chapter 4

Energy Options in Irrigation

CONTENTS

4.1 INTRODUCTION

'Irrigation' applies to the pumping of water, the supply of water and the supply of water to agriculture, where applicable. Water has to be drained from nearby channels, such as lakes, streams of water or canals, or from underground sources of open ponds and boreholes. Usually, on the surface, this water is drained into canals, lakes or higher tanks. The strength criteria for raising water is centered on the head, multiplied by the volume of water per cubic meter (vertical distance from the water supply to the ground in meters) (m3). The demand for energy will be higher than that indicated by this simple equation due to the friction and leakage of the distribution system (the dynamic head is the steady head or vertical height plus the additional height associated with the loss of friction and water withdrawal).

Figure 4.1: Irrigation involves the artificial application of water on crops.

Source: https://www.nicjamaica.com/types-of-irrigation/

First of all, the option of the model of the water supply scheme to be used is a crucial step in understanding the necessary pumping systems and the energy and power supply requirements. What irrigation method is this? Various factors are most important, including agriculture, temperature, location, size/area of agricultural production, the amount of water required over time, the cost of machinery, access to land, the technical potential of local farmers and the availability of machinery, including local maintenance, repair and replacement capacity. The highest and average demand for water differ and the conditions for water management (e.g. lakes, cisterns) vary from one irrigation system to another. More costly irrigation equipment can only be made available to targeted small and medium-sized farmers (farming on less than two hectares) by membership in manufacturing companies or cooperatives or farmers' unions where they have ample ability to collect capital and are able to provide funding for the installation, repair and replacement of the technology (Asif & Muneer, 2007).

4.2 IRRIGATION TECHNOLOGIES

Surface irrigation. Water passes by critical gravity forces in surface irrigation systems to moisturize and penetrate the soil. If irrigation allows fertile soils to flood or almost flood, this process is sometimes referred to as "flood irrigation" Traditionally, this method of irrigation has been the most commonly used. The levels are managed by a dike where the water level from the irrigation scheme can be regulated by the simple soil dams. In certain situations, the water in the ditches must be drained or pushed to the surface of the ground by human or animal forces. It has low costs and the advantages of fundamental technologies are offered. Disadvantages include heavy water consumption and possible high evaporation losses, as well as long-term issues arising from elevated soil salinity, which can result in routine floods due to this form of irrigation at certain locations. Sprinkler/ spreader irrigation: sprinkler or overhead irrigation requires piped water which is applied to one or more central places throughout the area through a high-pressure sprinkler or firearm. There are several schemes, including center pivot, hydro-wheel Potential work savings and more effective use of water than surface irrigation are positive. These devices can be costly and require the operation and supervision of scientific forces.

Drip irrigation: drop irrigation, also referred to as trickle irrigation refers to drop by drop of water immediately or near the root area of the plant. Irrigation: This strategy is an extremely water-efficient irrigation mechanism that eliminates evaporation and destruction. There is also a need for less chemical inputs for the overall, reduced and continuous use of water. Water and operating benefits are extended by up to two variables for each generator in the irrigation areas. Less water pressure and energy consumption are expected, relative to some other automated systems, but downhill pressure management is not straightforward. In particular, the effort needed to filter water for the elimination of particles that may obstruct the canals could be greater than that of other maintenance unit irrigation systems (Batterberry et al., 2010). Drip irrigation, particularly for high-end automatic equipment, can also be a very costly operation, although there are many very cheap model systems available (that require much more labor). Irrigation by sub-irrigation or separation: this method of irrigation carries water from the bottom of the soil to the root of the plant. It is possible to store the excess for reuse.

In areas with elevated water tables and for commercial greenhouse activities, sub-irrigation is used for field crops such as tomatoes, peppers and canes. In the lowlands and valleys of the river, these networks are primarily situated on permanent grasslands with a range of pumping stations, canals, weirs and gates constructed to regulate the level of the water table. In addition to potential labor cost savings, the levels of greenhouse irrigation and benefits, including the recovery of water and nutrients, have also improved. However, substantial early investment and technological sophistication are high. Manual irrigation: bucket systems or irrigation can have low requirements for machinery and equipment but high inputs for labor. This type of methodology is capable of simple farming, market farming or suburban farming. Usually, the size and expenditure of the technology would direct the decision on the irrigation technique to be used. Manual systems are planned for the smallest production plants with the lowest upstream and overhead costs. Small-scale agricultural pumps would draw interest (less than one ha). In addition to the scale of the production phase, the scope, costs and capabilities of the technology required are growing.

4.2.1 Pumping Technology

When selecting the energy resource to be utilized during irrigation, the pumping technology utilized is a major factor to consider. There is a wide range of technologies available for irrigation pumping, but not all of them are suitable for each irrigation system. Pumps vary in terms of pumping, size and capacity, type of water supply used (ground or surface water), size/area of irrigation needed, expense and technical difficulties, etc. Manual or motorized (mechanized): two large types of pumps are specified:

A variety of small-scale irrigation hand and foot pumps are available. They're both sucked, in whole or in part. Single-cylinder submersible friction pumps and hand and foot-powered rope-and-wash pumps are widely used manually, but their limited efficiency is usually limited to the supply of small-scale drinking water. When irrigation is required, a manual suction pump with 2 cylinders and a suction pump will reach a depth of 7-8 meters.

The output for two-person foot pumps is approximately 7.5 m3 per hour at depths of up to 5 meters, with outputs of up to 4 m3/h deeper at 7 meters. A hand-operated two-cylinder irrigation pump can produce from four to five m3/hr at the same depth, but its physical characteristics limit its time of operation, decreasing its efficient daily efficiency. In the 10-15 m3/hr range, the low lifting tray pumps with wide diameter tubes used to supply surface water or very low water wells of 2-3 m have a higher capacity (Badawy et al., 2010). When the weight of the user is used as a counterbalance and results from the superior force and strength of the lower body muscles compared to the upper part of the body, the foot-operated body pumps have higher outputs than the manual ones. In order to obtain higher daily costs, it is possible to use them for longer periods than hand-base pumps, which may be a safer option for the extensive irrigation needs of vegetable crops.

In fact, the centrifugal surface pump with a head cap of 6 meters is the most common engine pump used in small horticulture. This motor-powered mechanical pump is normally between 3.5 and 5 horsepower. Accessibility is 20-50 m3/h powered by petrol. The pumps range from \$100 to \$450. With respect to the fuel and power supply, the operating costs differ (e.g., an average of 0.4 L per hour of gasoline). On the other hand, diesel engine pumps typically have a capacity of at least 10 HP and thus have a higher power demand of around 2 L per hour compared to standard petrol engines used to power irrigation pumps. Higher mechanical centrifugal pumps are also available, which are costlier and have a higher capacity. The 5HP pump, built in India or China, costs between \$200 and more than \$700 and produces an average of 40m3/h and is dramatically different in terms of efficiency. The fuel consumption varies depending on the power, performance and pumping head of the engine. Surface wells and sources can yield several hundred cubic meters per hour from larger pumps, including those operated by vehicle engines.

The submersible electrical pumps mounted under the water table in the wells must be drained from higher concentrations in order to drive up the water by gravity. Occasionally, submersible pumps prefer to be used at depths greater than 10 meters and for broader irrigation purposes (greater than 4 hectares). A 3 HP submersible pump can reach a depth of almost 40 meters and a pump with a capacity of approximately 9 m3/hour. Motorized submersible pumps can be used to meet the requirements of farms of almost any scale. In most cases, the cost of submersible pumps is between \$600 and a few thousand dollars. Based on the requirements for the digging and finishing of wells and pipes, the extension of electrical cables or the building of non-grid distribution plants, the installation of reservoirs (tanks) and others, construction will result in substantially higher overall costs. On the basis of the well width, location and additional equipment requirements, the total cost of the whole system can be easily met or surpassed by thousands of dollars.

4.3 ENERGY SOURCES

Various pump systems can also be used as an energy supply and can be fuel-powered electric pumps, such as submersible pumps that can be operated by a manual lever, a merry-go-round pump and a gasoline-powered mechanical pump.

4.3.1 Fossil Fuels

The motorized pumps are used by fossil fuels to transfer power to the pump either by a power turbine or by a revolving belt and a vertical spinning shaft. Any submersible engine, such as piston pumps, operates by direct contrast movement (i.e. advancing cavity pumps). Pumps are much more expensive than centrifugal pumps, but they are often more efficient. Any of the pumps can also be used to power biofuels generated from different sources. Progress on the energy market is rapid and stable in terms of creativity and the use of clean energy (Cardenas et al., 2016).

4.3.2 Wind Energy

Wind energy is used to pump water more regularly, whether in small or large amounts, and to irrigate crops and/or livestock. In fact, for many centuries, the pumping method has been used to drain swamps and pump marine waters in Europe, South East Asia and China, particularly since the 13th century in the Netherlands. The so-called American agricultural windmill pump was the dominant type of pump used in the early 20th century. There was a windmill around the motor of the piston. The industry version, made of durable materials and with automatic structural controls and a working life of almost 20 years, is still commonly used, especially in the large Australian and Argentinian ranches. However, the expense is prohibitive. Other forms, which are essentially small workshop shapes, have also been refined. They have similar benefits throughout the service, even though they are not as efficient and can run independently. The pumps are included in this fact sheet in the following report. While they are less powerful than diesel or electric motor pumps, they do provide huge amounts of water pumping (5 to 25 cubic meters a day for most versions, but up to 100 cubic meters, their flows are completely dictated by the windmill scale, well depth and characteristics of the local wind). The primary use of windmill pumps is irrigation and feeding of livestock. Windmills can run when the wind speed exceeds 10 km/h, but only when the wind speed is between 15 and 20 km/h can they perform very well. When the winds rush at more than 40 km/h, they are not safe to use. Air saturation has a small effect on these values, which increases with a drop in air density and a rise in height (by around 1,000 meters). In areas where the wind blows at an average speed of less than 15 km/h, these water pumps are very similar to those of the engine pumps. Typically, this occurs in sparsely forested areas (savannas, desert and desert regions) or in more agricultural areas. First of all, there are two types of windmills: windmill water pumps and windmill motors.

As wind energy is used in both situations, working conditions are very similar, but wind turbines are being used increasingly all over the world as power sources are decreasing and emissions problems are escalating. It should be remembered, however, that, as is always the case, a windmill pump may be coupled with an electric generator to charge the batteries used: to supply an electric water pump on those days where there is insufficient wind or to provide some sort of energy, regardless of the kind of pump used in the well, for the daily needs of the farm or farmhouse. Windmill pumps transform kinetic energy directly from the wind into mechanical energy (Ceylan et al., 2004). They are a piston or a propeller powered by a rudder in the wind mounted on a shaft. They have a top frame with a return to the well-driven framework, a pump, sometimes submerged, and a water pressure outlet. While their function is simple, windmills, with the exception of simplified entry-level versions, are still very complex devices. The diameter of the windmill wheel (formerly sails) depends on the depth of the well and how far it is required to force the water to reach the flow rate. Building height and construction materials used are highly dependent on the possible wind strength and barriers in the area (none of which must be placed within a distance of at least 150 meters). Depending on the wind and the characteristics of the turbine, the flow rate of the windmill varies considerably. The volume would also vary from 200 L/h to 6 L/h to 7,000 L/h. Windmills may be designed with multiple attachments, such as a lever to stop them from the ground, an inlet ramp, a way to change the speed of the rotor, or even an automatic stop system when the wind force is high (e.g.

10 meters per second). That is why the choice and expense of a windmill depends largely on its characteristics, the pumping method to be used and the characteristics of the area in which it is mounted. This makes it easier to use many types of pumps. Instead of the shelf pumps, there are also more custom-made pumps available. This large range is further explained by the fact that the windmill consists of two key components (the windmill itself and the hydraulic system). As a consequence, the type of pump used by windmills also varies. Centrifugal drives and piston pumps are commonly used.

- *Piston pumps*

  This is one of the most widely used pump shapes. Due to the wave, the wheel on which the windmill blades are placed rotates. This wheel is mounted on a shaft whose revolving motion is transformed by means of an arm plate or a connecting rod into alternating vertical motions. The water pump piston below pushes the movements through a steel pin in this fashion. The pump itself is normally housed in the well. It is important that the smallest wind radiation be used because it would eventually require a very high torque to work. In certain cases, the windmill can have a very large number of blades. Usually, the number of blades is between 15 and 18. Even if the wind is relatively poor and basically decides its performance, the curve and the number of blades allow the windmill to start more or less quickly. In some cases, though, these things are less interesting than those using centrifugal pumps. This is since, due to the alternating pumping operation, the structure creates mechanical stress, resulting in faster wear (Cook, 2021). More special device experiments also show that only very limited speeds have replaced the optimum operating range for windmills and pumps. Power losses within this optimal range are thus very high and therefore require tight cooperation with skilled personnel.

- Centrifugal pump

  The torque is smaller than the column pumps. In this case, though, there must be a small number of blades in the windmill. The benefits of these pumps are also that they have an optimum operating range and can be adjusted to match the windmill, meaning that the wind pump performs better at a higher speed range. The performance of windmill pumps depends on a number of factors, but wind speed and pumping depth are the two major factors. The depth of the pump depends on the windmill being mounted. Highly decided is the pumping mode. If the windmill is mounted directly above the well, the suction or friction pump can be used, but if the well is offset, the suction pump can only be used. The windmill can be installed in an offset configuration up to 100 meters away from the well, but the suction height is only 7 meters high.

As the wind does not blow every day, depending on the usage of the well (irrigation, industry, stock, drinking water) or the output of the wind pumps, if a pump is required to service the battery, it should schedule the installation of an additional storage tank to supply water for a few days or install a new generator (for t). It is best to seek professional advice in order to acquire the experience of those who today need it locally. The power of the windmill should be balanced to the mode of flow from the well. For eg, a small or medium-sized windmill is not used to generate, whether it is high because it does not operate properly and is only effective if the wind is strong when there is a possibility of a break. It doesn't work correctly.

4.3.3 Diesel and Petrol Energy

There are two primary kinds of generators that include motors or power stations, and so on. Otherwise, gasoline, diesel or 1.p.g. Carbon engines; (1.p.g. - liquefied petroleum gas supplied in containers and typically mainly propane or butane) are the main fuels used to power these pump engines. The fuels are fed into the engine and compressed to form air. The air charge is immediately compressed and the engine ignites its fuel by heating effect; finely atomized droplets are sprayed into the cylinders by the injection nozzle when the temperature is sufficiently high, at the right time. On the other side, the ignition engines work by mixing sprayed fuel with oxygen, compressing the mixture and triggering it at the right time when the ignition plug is electrically discharged from the cylinder. In order to achieve the high pressures required for compression ignition, diesel engines must be heavier and lighter in design, and therefore a high-pressure metering and injection pump is needed to transfer the fuel at the right time using the injection nozzle. The diesel injection pump and the dust are designed to a high degree of accuracy and do not reflect expensive components when mass-produced (Cullen et al., 2011).

Oil/gasoline and kerosene generators are inevitably heavier, smaller and larger than petrol engines. In general, diesel engines are costlier to produce, but they compensate for this by being heavier, stronger and more efficient (but more complicated to maintain in good running order). First, increased compression allows diesel to 'breathe' quicker and to draw more air per stroke, regardless of the size of the combustion chamber; second, fuel injection enables diesel to run on a significantly lower fuel-air mixture than the comparable s.i. A spark-ignition engine should not be designed to run at such high compression pressure, otherwise, the "knocking" or "pinking" will allow the fuel/air mixture to fire prematurely. Another less common advantage of diesel is that it is 18 percent "richer" in terms of efficiency and durability per liter of energy than petrol (mainly because of its high density). As a result, diesel is usually chosen as a source of energy for a long period every day. However, the pumping of gasoline or diesel is mainly dependent on the pumping scale needed. In the event that a compact, lightweight, portable computer is needed and can only be used 1 or 2 hours a day and where simple maintenance is possible and reliability is essential, i.e. The farmer is only expected to pay a certain amount, so that the petrol or kerosene engine is the cheapest and is often used for this purpose.

4.3.4 Solar Powered Irrigation

Figure 4.2: Solar energy is among the latest trends in irrigation.

Source: https://futurepump.com/why-your-petrol-pump-costs-you-so-much-money/

Farmers have long played an important role in our society by supplying food to the people of this country. It should be remembered, however, that not only food is supplied, but electricity is also supplied, which is essential today, especially in terms of renewable energy. In practical terms, farmers can generate electricity from wind, solar or biomass and use it for their own plants. For farmers, solar energy may be one of the easiest ways to generate electricity. Farmers typically have a number of big homes, which are only below the sun and without being hidden by the shadows of the trees, making it a perfect place to mount a photovoltaic system. As a result, solar energy use in agriculture is becoming more widespread and renewable energy can be used either on the farm or on the local grid, which provides additional revenue for the farmer. Irrigation is one of the agricultural areas that benefit most from solar energy, especially in arid regions (Chowdhury et al., 2020). The key explanation is that with light, irrigation is achieved: as the sun rises, the watering system nourishes it; well when the sun shines a lot, we know that more water is needed for crops. And, when it's really needed, there's a massive amount of energy. The pumps used to store water are filled with solar cells. Using a generator that then feeds an electric motor that drives a pump, the solar energy used by the cells is converted into electrical energy. Many conventional pumping systems only run on diesel or central power grid. However, these two modes of service are disadvantageous relative to solar pumps. Connection to the electricity grid is not always guaranteed in many rural areas, especially in developed and emerging countries.

In this case, farmers cannot rely on conventional irrigation methods. Farmers will also guarantee stable supplies of power through the use of an autonomous and renewable energy system and prevent saturation of the public grid. Diesel pumps are less powerful than AC pumps, allowing for greater stability. One of the main drawbacks, though, is that it relies on the procurement of gasoline and increases the effect on the environment. Diesel pumps are cheaper than solar pumps, but the cost of service is very high and depends heavily on diesel costs. In solar-powered systems, however, this device is relatively costly and the electricity supply is free, but there are no direct running costs during the amortization cycle (only the maintenance costs must be considered). In the long term, solar pumps often represent green investments. The drip irrigation method is very easy to use. Water drips continually from the roots of the plant at frequent intervals with the aid of numerous pumps, tubing and hoses. A water sprinkler system that evaporates water into the air or invades soils where there are no trees, does not waste water when the waterfalls exactly as it needs to be. This helps more crops to grow with less water, making it a highly efficient irrigation system with drip irrigation technology. A compact solar panel with photovoltaic modules (up to 3kW) that can be drained from wells or reservoirs is attached to a wheeled pump in the Wien Energie solar irrigation system. You will measure the amount of electricity produced by your mobile app.

Then, through the tubing, the solar pump distributes water directly to the crops. This system can now be produced after a good 3.5-hectare evaluation in Guntramsdorf, Austria. Drip irrigation can also help to save water effectively in countries with high temperatures and limited water availability. This is all the more meaningful because the farmers face three challenges: maintaining water, saving money and saving energy. Handheld solar drip irrigation systems would be the ideal way to solve these problems (De Andrade et al., 2015). Although these initiatives appear to be very costly to tackle, a variety of R&D programs are working to democratize the use of solar energy in agriculture, which could play a key role in solving food and energy shortages in the future (and even now).

4.3.5 Electrically Powered Irrigation

Electricity, when readily available, is the cheapest source of irrigation electricity. Electricity has many advantages for agriculture. The dumping of liquid fuel in the area of surface water wells presents future risks of contamination eliminated by the use of energy. Electrically driven pumping plants can be operated more effectively remotely and assisted by a technologically sophisticated irrigation administrator. Further irrigation is motivated by higher initial pumping costs and higher overall combustion engine running costs with the use of electrical fuel. Many irrigators argue about the degree to which electricity is provided by the irrigation scheme. The response depends on your location, the service provider and the energy available at the moment. Almost all farms have one-phase electrical capacity. Three-phase field-scale irrigation systems, electrical motors and controls are designed. Electricity converter or variable-frequency drive systems offered by the farm are suitable for producing electricity from a single-phase line in three stages, but would mainly be limited to small engines to alleviate problems with the brownout on the service line. Electric utility utilities should also allow farms to buy and run soft start pump engines and variable frequency drives to reduce start-up loads on larger power pumps, according to the Michigan State University Extension.

To fix the problems of browsing. The energy supply from the electricity supplier to the pumping site is the shortest and most productive source. The maintenance of irrigation systems requires a minimum quantity of farm machinery. Power service companies can also provide a calculation of the cost of connecting three phases of electrical energy supply to your home. In certain cases, the projection will be free if the sum of energy would be required in the future. More efficiently, a quick reduction from a new collection of transformers on the same three-stage line could be between $500 and $2,000. If the 3-phase system does not close, the construction cost is expected to range from $10,000 per mile to $30,000 per mile. These costs must be paid by the prospective irrigator in order to boost agricultural operations. Irrigators may also calculate initial equipment benefits and annual repair and energy prices for power for combustion engines for a period of 5 to 10 years. For a typical 160 acre irrigated area, the upgrade of the power grid in that district would be covered by a 10-year savings of $25,000 or more. In addition, in the next two years, most power producers have a scheme to repay the original investor with a share of the price, as other consumers connect to the new three-phase line (Delmastro & Gargiulo, 2020).

Chapter

5

Energy Options in Harvesting Agricultural Production

CONTENTS

5.1 INTRODUCTION

Equipment supply chains are the availability of the necessary equipment, replacement components, service experience and fuel in the normal consumer travel portfolio is a prerequisite for sustainable technology transfer. For local agricultural producers, these products must be affordable. These requirements may not be fulfilled when equipment is supplied and paid to non-governmental organizations including NGOs, bilateral aid organizations, UN institutions and other sponsors of projects. Stakeholders and promoters should collaborate to ensure a secure local equipment supply (whether manufactured or imported), replacement parts, machinery, lubricants, and local language user guides to prevent interruptions of agricultural production or processing caused by interruptions of the supply chain or lack of knowledge on the necessary technology. Sustainable availability of technology ensures that a chain of traders provides distribution points from the point of production to the location that consumers buy technologies (Edelenbosch et al., 2020).

The ultimate sales price contains manufacturing costs, shipping costs, trade margins, taxes and other fees included in selling the commodity to the consumer. In addition, appliances and components bought must have admissible after sales guarantees and repair/service agreements. If donor organizations do not make adequate arrangements and supply only hardware, resource needs can be significantly jeopardized. Besides, where numerous benefactors give gear, the host country government can strive to coordinate giver uphold and build up the help of just one or a couple of the key hardware providers to decrease the requirement for a few manuals, different instruments to oversee, fix, supplant and keep up the intricacy of numerous new parts. In this area, the absence of coordination makes significant difficulties for proceeding with activity and upkeep of equipped, for example, solar-based antibody fridges and little diesel generators for off-matrix power age all through the created world.

It is critical that many of the horticultural machinery comes with the local instruction manuals (s). In donor machinery procurement structures, this essential element of sustainability is sometimes absent. To repair missing or worn-out pieces, equipment materials need to be readily available. The end-user and contractors should recognize the replacement parts and tools needed to fix key equipment quickly. In certain cases, even when the equipment is imported, components could be produced locally. If improvements need advanced know-how, skilled local mechanics are a key link in the supply chain of engineering. Repair facilities and substitute costs can be covered by repair arrangements, either by customers or by equipment manufacturers.

Figure 5.1: Energy is needed on the whole stages in the agri-food chain: in cows, fish, yield, and timberland items creation; in post-collect exercises; in food stockpiling and handling and in food transportation and dispersion; and in food readiness. Harvesting and transportation of produce require a large amount of energy.

Source: http://www.fao.org/e-agriculture/blog/energy-smart-agriculture

It must consider both high oil fuel prices and possible difficulties in fuel logistics. Liquid fuels may never be commonly accessible in remote areas but may need to be shipped from the closest sales points land or by ship in limited amounts. This also greatly raises the price of fuel charged by agricultural farmers. Fuel supplies are unreliable in many rural areas, which can lead to broad and/or unexpected slow periods. In the case of such disturbances, like watering of the fresh-picked crop or cooling, at crucial points in the manufacturing and sales cycle, production may be lost during the year.

Nearby energy choices are regularly prescribed to be utilized in regions with lacking fuel supplies. These assets incorporate siphoning, transport, and different pieces of the agrarian cycle human as well as animal energy, nearby biofuels, for example, jatropha-curcas oil as an incomplete or complete supplanting of charcoal, wood, diesel fuel, or farming deposits (for example, nutshells) for powering generators and boilers; and miniature wind, wind, and sun-based power. In several rural areas, spoilage of fuel is a common issue, either accidentally or deliberately (Elghawi & El-Osta, 2015). Unofficial containers like plastic jugs are prone to be polluted with petrol, which is supplied by water and soil. Filtration is crucial in eliminating solid particles, but it is hard to eliminate any surplus water. As a result, engines cannot perform effectively, and mechanical disruptions may become more frequent. In certain places, grid electricity may be accessible and secure. The existence of a local grid does not, however, mean the electricity supply either is accessible or quality.

If frequency or voltage regulation is weak, spikes and surges can easily kill electronics and motors. For horticulture production, the use of a combination of propane and diesel generators, solar or wind generation, bio-powered gensets and micro-hydropower can be more cost-efficient in areas with inadequate electricity supplies. Adding continuous energy supplies as a cost-effective alternative is strongly recommended for limited, high-value uses of electricity such as electronic, telecommunications, and computer control. The cost effects of low electricity quality and grid supply failures will lead suppliers to implement generators onsite when evaluating the maximum relative prices of different electricity resources even if the electric bill for onsite energy units is significantly above the national grid price.

Other than dependable hardware and energy supplies, agricultural creation should be supportable monetarily. The accompanying techniques are suggested that the interests in agriculture ought to be deliberately considered as a necessary arranging step:

- The innovation to be received ought to be the base size to cover the present and extended requirements of an endeavor (i.e., proper scale)
- In terms of buys, repairs, and regular activities the technology must be the most economical alternative to meet the requirements of a company.
- The expensed cost and repetitive costs of maintenance and service should be offset by the cumulative revenue produced.

When all is said in done, overabundance limit and energy supply add to abundance costs and diminished financial practicality. For example, a little green movement may be more monetary in fulfilling its water supply necessity through numerous Manual Pumps than a mechanized siphon if a mechanized siphon has a similar limit as four manual siphons. For the existence of the advances viable a money-saving advantage investigation ought to be made to pick the most reasonable other option. Contender advances, for example, kWh/MT or $/MT of flooded, passed on, cooled or bundled merchandise ought to be comparable on broad standing.

Contrasted with lower get and energy costs and higher work costs in a lower innovation variant, compromises can involve the expanded expense of acquisition of similarly high-innovation hardware, given their relative presentation. Little energy cooling techniques after harvest can bring about more noteworthy misfortunes than mechanical cooling, yet energy reserve funds can more than counterbalance these misfortunes. Higher fuel costs will drive exchange energy/work toward more work serious, less-innovation exercises. No financial maintainability ensures even the determination of the most reasonable least expense innovation per unit creation. If the outcome is yet negative, conditions for business green creation in each area with setting up provisions of info, market access, maker cost and other creation factors are not great (Ersel et al., 2004).

Figure 5.2: Many energy requirements of agriculture are not natural to sustainable or decentralized energy. In most African countries, solar is the lowest-cost electricity. Many farm systems however contain similar large quantities of power at rather unusual or even regular intervals.

Source: https://cen.acs.org/food/agriculture/Protecting-harvest/98/i6

5.2 HARVEST AND POST-HARVEST OPERATIONS

The workplace might be latently vanishing by soaking the dividers of a packing house or by utilizing pore materials toward one side of the construction, (for example, straw cushions wet with water, as generally found in huge nurseries). This straightforward strategy will consider dynamic cooling of the work regions through water dissipation if the air is gotten by low-speed ventilation providers through the wet cushion or divider. In dry environments, yet less so in high stickiness areas, this can be exceptionally proficient.

5.2.1 Electric and Fuel-Powered Technology

Heated air treatment: if the heated air is utilized to treat onions, garlic or other bulbous cultures, the treatment time is suggested at a temperature of 35°C to 45°C and relative humidity (RH) of 60 to 75%. Through harvest, power needs will rely mainly on the environmental temperature that dictates how often heat is needed to attain the temperature needed for cure. For example, for the treatment of yams, a most extreme temperature of 30°C is prescribed for a time of four to seven days and RH 90%. Lotions are accessible available, however moisturizing the floor will keep the HR high (Esen, 2016). For warming of 1 MT of yams of 15 °C to 30% C in 24 hours [31 kWh-th per metric ton], a radiator limits of 4.540 BTU each hour (1.3 kWh) is required. To keep the potatoes at 30°C, a radiator isn't needed. After the harvest is heated up, the air temperature can gradually return. After the yield is warmed, the encompassing temperature may return gradually. At the point when the encompassing temperature during harvest is fundamentally above 15°C, the energy utilization is diminished generously. Notwithstanding, cooling ought to be utilized when the encompassing temperature is more than 30°C on the grounds that relieving will not occur over 30°C.

Washing and cleaning: Mechanical washers utilize conveyors, air dryers, roller brushes, and sprayers to move produce through a pressing line. Conveyors typically have engines of 0.5 - 2 hp (needing 300 watts to 1.5 kW of electric force). Drying should be possible utilizing air blades and brushes with flipper bars, needing unheated air and basic transport frameworks.

Drying and waxing: Machines for washing/drying incorporate water sprayers, air warmers (electric, flammable gas and propane, temperature change) and air knife blower (air blade), with 4 to 10 hp power necessities (3-7.5 kWe). The utilization of power dependent on orange and lemon is from 1 to 2 kWh/MT for drying, waxing, and washing though gaseous petrol consumes somewhere in the range of 60 and 90 MJ/MT for pressing. The utilization of dissolvable-based wax plans in numerous nations that frequently do not require warmed air to dry wax might be legitimate.

Management of pests and diseases: Plunges can be warmed by wood or biomass fire, propane, gaseous petrol, electrical or sun-based water radiators. Boiling water treatment for the suggested temperatures (commonly from 40 ° C to 52 ° C for 2-5 minutes, or 400 L of water from 20 °C to 25 °C, or 9,2 to 14,7 kilowatt long stretches of power, contingent upon the kind of material being taken care of, is followed by ice showers to decrease the temperature quickly. Temperature necessities vary by the item being dealt with, yet the standard reach is 40 °C to 52 °C. Boiling water plunges can be warmed utilizing propane or power for more even controlling of temperature.

Grading and sorting: Arranging and reviewing can be quickened by mechanical measuring utilizing transport frameworks. Veering bar sizers and belt sizers are normally intended to incorporate 1 to 2 hp engines requiring 0,75 to1,5 kW of power and significantly other essential transport structures size machines. Power utilization for arranging/measuring, considering accessible information for lemon and orange packing, ranges 0,8 to 1,7 kWh/MT (Ghalehkhondabi et al., 2017).

Packaging: For mass bundles of less sensitive merchandise, for example, carrots or potatoes, programmed packing by gauging and sacking is generally gainful. Power use in the pressing of lemons and orange packing is somewhere in the range of 0,7 and 2.2 kWh/MT.

5.2.2 Technology Powered by Renewable Energy.

Solar water heaters will significantly minimize the demand for electricity or propane, reducing the fuel demand by 80% to 90%. Solar collectors: The single glazed, single-panel, single-glazed, flat plate solar collectors can be manufactured locally, and are commercially available worldwide. Solar collectors capture solar electricity and transmit the resultant heat to the water or air. A functioning solar- based warming framework is a solar-powered gatherer. Solar- based systems take in sun-oriented energy and move the subsequent warmth to water or air. The level plate authorities for low-temperature water or air warming applications are the most well-known kind of solar- powered gatherer. An encased box (plastic, metal, wood, and so forth) with a reasonable glass or plastic cover and a dark safeguard plate is utilized for a standard level solar- powered plating gatherer (Grunewald & Diakonova, 2018). These pickers are valuable for the creation at temperatures of up to 80°C with warm water or air.

The measure of helpful warmth that a solar- powered gatherer can emanate (as sweltering or blistering air) relies upon the strength of occurrence sun-based radiation, the daylight reflected and retained through the straightforward surface, the percent of the energy that is communicated from the assortment surface and the distinction in temperature between the approaching (water or air) and heat transfer liquid. The capacity in the huge outside tank, dark and purposefully left presented to the sole, of the water utilized for boiling water plunges will raise natural water temperature by 30°C or more and lower resulting heat costs by up to 30%.

Figure 5.3: Harvesting is by far the most labor-consuming operation in the planting season for smaller farms with limited automation. The costliest and specialized farm equipment such as the combine harvester is used for the harvesting of vast mechanized farms.

Source: https://en.wikipedia.org/wiki/Harvest

Water heating integral collector storage: The boiling water stockpiling tank is the sunlight-based safeguard in the Integral Collector Storage (ICS) unit. The tanks or tanks are mounted on one side in a secluded glass box and painted dark. The sun glimmers in the coating and hits the dim tank and warms the tank's water. There are a few adaptations that are fitted with a solitary enormous tank (100 to 200 liters), while others are furnished with metal cylinders plumbed in number (100 to 200 liters of in general capacity). These gatherers weigh 125 - 200kg when full, so any place they have been mounted, the design should be sufficiently able to convey this critical weight. The single tanks are commonly made of steel. Single coated gatherers, for example, this has productivity of under 50°C for low-temperature applications. Higher temperatures request double vitrification and more prominent gatherer protection (Goiri et al., 2015).

Fans powered by solar energy: Solar- oriented force providers: Ventilation in a structure's rooftop will fundamentally lessen the warmth development consistently and sunlight-based energy suppliers give sufficient air course to utilizing an evaporative cooler during cool, little stockpiling. For level just as pitched depressions, solar-based fans (counting solar-oriented boards of 10 to 20 watts) are sold for a retail estimation of from 400 to 600 dollars. The rate of fumes goes from CFM 700 to CFM 1400.

5.3 COOLING

Cooling in the post-harvest processing chain is one of the most critical steps. Lowering product temperatures after harvest can reduce breathability, prolong shelf life, and maintain quality significantly while minimizing volume loss by lowering the water loss rate and decrease. This first cooling phase is commonly alluded to as a pre-cooling process as the product is put in the cold or loaded in cooled vehicles or marine vessels as early as feasible after harvest and prior to production. The time span of usability of the post-reap item relies upon keeping cool produce. The degradation rate rises essentially as the temperatures of the item rise. For example, the pace of spoilage can twofold or even triple with a temperature ascent of 10°C, which implies a lot more limited life after gather.

5.3.1 Low-Tech/Manual Technologies

Shading: Covering and protecting new produce from direct daylight is a modest method of limiting warm utilization. To give profound shadow over all gathering focuses and work zones it is prescribed to utilize material or texture rising. A profound rooftop development (at any rate one meter) can conceal the windows or entryways and a light-hued or intelligent rooftop can decrease under the haven surface temperatures and temperatures to up to 20 °C.

Coldwater (from mountain streams or deep wells): All things considered, in many districts of the planet, water is in every case a lot cooler than the air temperature. Generally, water from profound pools and mountain streams is estimated at a normal yearly air temperature in the district. Well water might be utilized for hydro-cooling and as a shower of mist in the capacity climate to hold high relative stickiness. Be that as it may, water from streams is likewise contaminated and is not suitable for food correspondence.

Passive evaporating cooling: At the point when air is gotten through the wet cushion, ventilation fans may give latent cooling by water dissipation from the divider by wetting the mass of a packinghouse or by utilizing permeable material toward one side, (for example, found in enormous greenhouses).

5.3.2 Electric and Fossil Fuel-Powered Technologies

Cooling of rooms: Room cooling is an exceptionally speedy, however lethargic approach to diminish the item's temperature before chilly stockpiling, where the item bundles are inside a virus room and gradually invigorated. Room cooling commonly takes 24 to 48 hours or more and for profoundly short-lived crops is not suggested (Hamlet et al., 2010). During the underlying cooling, fans ought to have the option to supply 90 CFM/MT. When target temperatures are met, fan speed can be diminished to give the wind current of 18 to 25 CFM/MT.

Ice: A central ice processing plant and ice delivery system provide for cooling where there is no energy or mechanical cooling. This was the initial justification for the growth of the long-haul company in the USA. However, ice-cooling is very ineffective since the product is cooled utilizing only around half of the cooling effect. The remainder is lost to the warm atmosphere. Furthermore, substantial ice leakage will occur as it melts into the cooling system in transit from the main cooling facility. Except in an isolated setting (for example an ice chest container), if containers and ice could be utilized at least 50 percent of the initial ice will be wasted before it is utilized. As an ice maker has many elements powered by electricity, ice production is energy-intensive and can be costly. The majority of the ice machines produce unadulterated ice per kWh somewhere in the range of 5 and 12.5 kg. The Crytec Bubble SlurryTM ice machine Modelle CR-004 produces 18.5 kg unadulterated ice per kWh with a limit of 146 kg unadulterated ice each hour, which is quite possibly the most energy proficient ice creators accessible. Since the ice requires 330 kg of new items to cool one MT by 28°C for a limit of 66 kWh, it is regularly restrictive to make frosts due to high costs.

Forced evaporative air cooling: The cooling phase is accelerated using an electric supplier and a water pad to transport cool air into containers of fresh produce. Manufacturing temperatures can be cut to a couple of degrees over the surrounding temperature of the dew point by evaporative cooling (the temperature at which dampness starts to frame on a smooth surface, demonstrating 100% immersion of the air with dampness). Air flux of 1 L/sec/kg vs a large variety of static pressure should be given by the fan. Doubling the airflow may accelerate a bit, yet the cost is substantially higher, as the fan requires four times more power to perform the identical function. The cooling impact of one kg of ice is around 316 BTU. The temperature of fresh products or water of around 28°C would be lowered by 3kg of ice.

Figure 5.4: The utilization of battery power for farming and equipment hopes to revolutionize farming by reducing cost and increasing efficiency. The agricultural face changes from large tractors driven by batteries to unmanned electric robots to batteries and solar energy.

Source: https://www.ee.co.za/article/electric-powered-farm-vehicles-set-to-revolutionise-agriculture-sector.html

Forced pre-cooling of air in a cold room: Cooling of constrained air (FA) will quicken cooling the readied product group from at least two days to under 8 hours stacked into a cold room. At the point when a cold room has the appropriate cooling ability, it raises the fan's force utilization by only 800 to 1500 watts an hour by introducing a versatile air-cooling burrow that can cool 4 cushions all at once. A 5-ton cold room cools 3 MT agricultural produce in 6 to 8 hours, from an underlying temperature of 27 °C to an objective temperature of 2 °C.

To handle the ventilation without undue pressure falling throughout the vessel, the winding region on the edges of the generated vessels must exceed at least 5% of the volume of the vessel. Fans for FA coolers normally work within a range between 0,5 and 2,0 liters per kilogram per second (1 liter/kilogram per second is equivalent to around 1 cfm per lb.). Doubling airflow would induce some cooling (maybe by 40 percent), but the cost of energy will increase dramatically as the fan will have to utilize 5 or 6 times the energy (Hossain et al., 2017). For example, 3MT wind stream at 1 L/kg/sec and 1.3 cm w.c. (pressure water section) needs 1.12 HP (0.85 kW). The size of the fan would develop to around 7 HP when the wind stream is multiplied. For most little cooling applications, diffusive fans with sent cutting edges are fitting. Modern propeller fans, universally accessible, are most appropriate for low pneumatic force applications. In the United States, the expenses are for the most part between $1,600 for a 1 HP fan and $1,000 for a 1⁄2 HP fan.

Early discoveries show that standard business constrained air cooling has an occasionally normal energy utilization of 55 kWh/MT with a gauge of 22-27 kWh/MT, in a power use study on coolers presently being led for the California Energies Commission. This requires some pre-shipment momentary capacity. Beneficial activities will run at about a large portion of the normal electricity per MT. Item execution is more compelling than some other factor during administration, as kWh is determined per box (Thompson and Singh, in press).

Hydro-cooler: Utilizing ice or mechanical cooling, the water utilized for cooling should be kept freezing. Water moves far superior to air, so cool hydro-coolers produce a lot quicker than coolers. Small widths, for example, cherries chill off in under 10 minutes in very much fabricated hydro-coolers. In 45 minutes to an hour, things of incredible breadth, for example, melons cool. Batch-style hydro coolers place over the tops of the packed containers one or more pallets of items and shower cold water, helping them to filter through the containers and touch the goods, and eliminating the heat that passes through the load. The door is normally open, but the door is closed during service to minimize heat and water losses.

5.3.3 Energy Consumption

In 2012, almost 800 trillion British thermal units (Btu) were used in the U.S. agriculture industry, or almost as much primary energy as the whole of Utah. Agricultural energy use requires energy for planting, harvesting and livestock production. Crop businesses use far more energy than animals, and crop energy is a greater proportion of the cost of farm operations. Indirect and direct use of energy in agriculture is included. Direct energy use includes the use for farm operations of renewables, gas, propane, electricity, and diesel (Huang et al., 2011). Indirect energy use involves the consumption in the fabrication of agrochemicals including pesticides and fertilizers of fuel and feedstock (especially natural gas).

Energy accounts for a considerable proportion of operating cost for most crops, particularly in view of indirect fertilizer energy consumption, as the manufacture of fertilizers is extremely powerful and needs large quantities of natural gas. Combined energy and fertilizers spending on certain crops, such as barley, wheat, maize, and oats, is over half of all operating costs. The ratio of direct to indirect energy consumption varies according to crop. For example, maize, also a source of energy for the processing of ethanol has comparatively little direct fuel expense but the largest proportion of fertilizer consumption.

In the livestock sector, the fuel expended is almost exclusively direct and comparatively small, as a proportion of overall operational expenses and overall, relative to crop activities. Livestock companies use direct energy for ventilation, cooling, lighting, warming, watering systems, engines, and waste management, while crop operations use vine, harvesting, irrigation and dry crops for energy consumption. In the study of livestock energy use, energy used in the processing of animal feed is not included.

Distillate fuel is the essential fuel for both animals and harvest activities for direct energy use. The distillate is utilized to drive crops, collect, control of weeds and other large equipment activities. Harvest drying is another fuel-creating activity and relying upon the yield type and its dampness level, the measure of fuel utilized is extraordinary. Power or propane is utilized in high-temperature dryers. Water supply can likewise be an amazing assignment. Albeit a few farms approach public water supplies, most ranches siphon water from sources and sources of groundwater. Most power is utilized to siphon; however, siphons may utilize diesel or propane at distant destinations (Liu et al., 2019).

Agricultural chemicals comprise pesticides and fertilizers, which is partially a bulk chemical sector. Nitrogen (ammonia-based) fertilizers are a source of heat and power that demand large quantities of natural gas as a feedstuff. The 2010 Manufacturing Energy Consumption Survey of the EIA reports that more than 200 billion BTU of natural gas as feedstock were consumed in 2010 and 152 billion BTU for electricity and heat was used by the US nitrogen fertilizer sector. Besides the fact that some farms are big energy users, they use renewable resources to generate energy. The most popular on-farm renewables are photovoltaics, methane digesters, and wind turbines. Renewable energy may contribute to compensating for the power obtained the renewables generated by farms are, in some cases, sold to distributors of energy and provide farmers with incremental revenue.

5.4 TRANSPORT

In the agricultural chain, transport happens a few times. Transportation from the farm to the warehouse and from the warehouse to the port (for export), or the nearby retail market (for local consumption). Further ground, air, or ocean transport will be required for sends out. Manual vehicle advances comprise essentially vehicles, trucks, or carts driven by creatures. By examination, a wide assortment of methods of transport incorporates non-renewable energy source fueled innovations.

Figure 5.5: An especially important segment of transportation is the harvesting and post-harvest processing chain. If the damages are so high, it is worth considering what can be changed from the time of harvest to the moment of market access.

Source: https://www.myfarminfo.com/blog/tag/crop-transport/

Customary vehicles can hold versatile coolers or protected boxes fitted with climate control systems and diesel generators (get trucks, flatbed trailers). A little disengaged cabinet (3,5 m3) with a creation of around

700 kg can be pulled up (USDA Porta-cooler 1993) as a trailer and can be stacked onto a get truck bed (10,000 to 12,000 BTU) with room cooling and diesel-controlled generator (2 kW). The utilization of plastic liners in compartments will limit water spoilage.

These units can be utilized viably for the transportation of tropical and sub-tropical plant crops at or above 10°C with great outcomes. Room forced air systems are not prepared to work proficiently at exceptionally low temperatures and to create low relative dampness, cooled air that can bring about unsatisfactory water misfortune rates during products of the soil cooling (Liu et al., 2013). Also, ice will develop on the twists and the climate control systems won't work as arranged at temperatures beneath 10°C.

The evaporative cooler is a significantly less expensive and more successful decision for the versatile cooler by and large. A profound cycle capacitor can control the cooling unit in the plan to run a little siphon (11 per minute, 10 watts or less) and 100 to 300-watt fans to move air into the cooler's wet cushion. The more profound battery supplies capacity to the water cushion. Significantly more force can be given. The inside of these cool boxes is 3.5 to 7.0 square cubic meters, and the fan should exchange air each moment with a static pressing factor of 0.6 cm w.c. (64 to 128 CFM roughly).

At the back of the pile, a fan ought to be given with the goal that the air, which is evaporative cooled, is completely stacked inside the detached box. If the unit is on a truck, it tends to be fitted over the truck using an air scoop for the pushing of air through the unit while the truck is in motion, so the fan power is needed just when the unit is fixed. The most mainstream cold trucks are 12 ft and 20 ft reefer trucks containing diesel generators for controlling the cooling framework. The reefer trucks are the most well-known ones. The trucks can contain up to six beds or ten beds that help 800 kg of items for each piece. As the cooling framework mounted on a virus truck is relatively little contrasted with the size of the motor of the vehicle, it is required to prompt an expanded fuel utilization of the general energy utilization of the vehicle. A reasonable gauge is that 5% to 10% more fuel will be utilized to move a similar burden on a similar distance in a refrigerated vehicle.

ThermoKing gauges that marine compartments or multi-purpose reefer holders need about 5.0 and 6.5 kW each hour of power for their cooling units (through modules on the boat and in ports during delays prior to stacking or get). New models can be substantially more remarkable and just require 3,6 to 4,0 kW to cool similar burdens. Around 80% of Europe's reefers have an electric champion which empowers the gadget to be connected to electric force by up to multiple times the energy productivity that can be accomplished by the utilization of diesel energy alone. For the operation of the generator unit on standard 20-foot to 40-foot reefer containers with a total rate of consumption of fuel based on the fixed point, estimates of 2.2 to 5.0 liters of fuel per hour are needed. Air carriers are not normally fitted with cooling systems. The energy usage for various transport modes is remarkably rarely compared. The most detailed research on non-cooled transport from Europe shows the energy use for different transportation methods (Liuyijuna, 2011). Small highway automobiles are far less energy expensive than large trucks, especially cars used for transporting small quantities of produce. The typical energy-efficient transit service is valued at around three times that of lorries, making it especially appealing as the prices of fuel are rising. The most effective means of transportation is marine transport.

Chapter 6

Energy for Drying of Agricultural Production

CONTENTS

6.1 INTRODUCTION

Recent technological advances have led to the creation of advanced drying processes, especially for numerous consumer and industrial applications and processes. Although modern drying techniques are very helpful in ensuring very good sanitary and efficient care, continuing energy dependency on the use of modern dryers is a significant and unavoidable expense. Drying can take place at high temperatures, such as fire drying or dielectric heating, or at low temperatures, such as at room temperatures for cooling and drying. In view of the strong demand for power and the inefficient use of traditional resources, such as fossil fuels, green drying operations using renewable energy sources must remain more effective and productive.

Drying is an effective way to prevent degradation by extracting enough plant moisture to preserve agricultural products for potential use, as the moisture content of certain forms of agricultural products allows them to die faster, especially in the tropics. In addition, the savings of agricultural products are feasible and cost-effective, as solar heating can be used in conventional open-air drying or in specially built solar heat dryers to avoid post-harvest harm typical to many developed nations. The method of using solar fuel inefficiently and therefore quite polluting, resulting in poor dry material quality, is used in many parts of the world where traditional open-air drying is carried out.

Extensive attempts have been made to enable the drying of various varieties of agricultural products by the efficient use of solar thermal energy. These plans involve a sheet with a drying panel that collects the solar energy streaming through the dryer and directly heats the drying panel. They have a low air passage, which means that no optimal drying temperature can be achieved on cloudy days (Majidi, 2014).

Unfortunately, the solar collector increases the passage of heat and air across many layers of trays into the drying container and also leads to irregular food-drying and molding by lowering the flow of air through several layers. In Bangladesh, Mohsin has studied the future and efficiency of various Solar Dryer designs, including natural solar convection dryers and forced solar convection. Mohsin argued that drying air temperature can be reached by adjusting the number of fans in operation by simply increasing the length of the collector or the amount of airflow (in the solar tunnel driver). Various researchers have developed solar energy boom systems, in particular agricultural products and places, including the development of solar-powered models in Cambodia for selected fish species, the determination of banana thickness cutting (*Musa* Spp). The only downside to a fixed solar dryer was that a lot of solar energy was wasted a few hours a day after the Sun's rotation left the perfect place to eat it. The visible decline is the integration of the solar surveillance system into the solar oil drying process. Maximizing energy production by means of a solar monitoring device is the key approach.

This is especially important, considering that solar thermal and photovoltaic systems produce energy that is directly proportional to the solar energy consumption of the unit, and it is important to improve the efficiency and reliability of solar energy systems that are commonly used in domestic and industrial applications.

In the past, renewable energy technologies such as solar panels have usually been positioned around 30 degrees south from east to west to maximize power production. Technical developments have led to the design and application of solar panels, mirrors or other sun-facing optical instruments.

This makes it easier to increase energy production, as the location of the Sun in the sky varies with the time of day and the seasons of the year, the storage devices are specifically aligned with the Sun (Mortazavi et al., 2019). A significant number of solar monitoring instruments have been used in the field of solar energy generation. Solar sun-drying techniques are rarely used. The goal of the research is to investigate the effects of solar surveillance on the generation of solar panel dryers in relation to the position of solar panels and open-air dryers.

6.2 SOLAR DRYING

Sun drying is a common drying method, as are free and sustainable sources of energy used in the production of fruit and vegetables, food, fish, beef, wood or other agricultural products. The method of sun drying is normal. A wider range of existing restrictions on sun-drying, including livestock disruptions, birds and rodents, quality deficiencies due to direct exposure to sunshine, dew or rain, soil pollution, dust and debris, are recognized for large-scale growth. Crops may be affected by domestic attacks at night and in poor weather, and they are both time-consuming and labor-consuming. The uniform drying process also enables the infestation and growth of insects and microorganisms. Solar drying is a drying mechanism for products with a closed system of elevated temperatures. It's very helpful to protect yourself from insects, bugs, rain or pollen.

In recent years, significant attempts have been made to use solar energy to dry it, in particular, to save crops and to take advantage of solar energy. Sun-drying is the most common food storage technology in most parts of India and the world, as solar irradiance is high most of the year. This strategy is more suitable for smallholder farmers with a shortage of energy during the day, who are unable to afford it or other drying fuels. The external bio-fuel heater can be used for drying at night or in poor weather conditions (Moran et al., 2014). Only the main agriculture sectors in developing countries can make high-temperature dryers used in the developed world economically viable and are usually inefficient for medium-sized and small companies due to their very high expense and organizational uncertainty. This provides a potential alternative to minimize post-harvest losses by cost-saving solar dryers and solar dryers for locally operated large plants. Farmers are encouraged to adjust economic conditions in order to produce high-quality, marketable products. In comparison to the low incomes of rural citizens in developing countries, relatively high initial solar dry spending remains a barrier to universal access to electricity. However, it would be economically affordable for farmers if it is made from local products such as wood, glass, etc.

6.2.1 Solar Energy

Light is the primary energy source for our solar system. Continuous nuclear fusion produces more energy in the very heart; a tiny amount in the sun is touching our world and bringing life to the world. Solar radiation is responsible for life-critical processes and motions, including temperatures, winds, sea waves and photosynthesis. From the beginning, solar energy was a staple of the global energy industry. All fossil fuels have been converted to solar (coal, gas, oil). The solar disk temperature of the sun is equal to radiation energy of between 70,000 and 80,000 kW/m2. The surface contains only a small amount of energy. In consideration of the fact that there is about 2 solar radiation in a year? Global electricity consumption is 1017 kWh, 10,000 times as high. Solar volumes are about 1,360 W/m2, minus the atmosphere (solar constant). Power is partly lost due to sunlight, which means that 800 – 1000 W/m2 (global radiation) is produced in the field on a sunny day during the day (Muralitharan et al., 2018).

Figure 6.1: Solar Drying.

Source: https://foodtech.uonbi.ac.ke/node/229

6.2.2 Global Radiation

Timing and intensity depend on the time and intensity of the sun, the variations in the atmosphere and, inevitably, the geographical environment. The horizontal surface of the Sun Belt region will also have a global annual exposure of 2200 kWh/m2. Global radiation generation was based on direct and diffuse radiation. Direct sunshine is part of the path of the sun. As direct solar rays and particulate matter are distributed in both directions, i.e. radiation, the dispersed radiation portion is released by different molecules. Dependent on temperature and spatial conditions, the volume of radiant light is diffuse. Clouds, atmospheric patterns (e.g. clouds and powder layers) and emission period in the atmosphere have a significant effect on the radiation proportion as well as on world radiation.

6.2.3 Drying

The method of thermally separating water content for the disposal of solids is also called drying. The moisture associated with this is water that exerts a vapor pressure below that of a pure liquid trapped in a loose chemical mixture, contained in a product matrix or even caught in a solid microstructure. It is known that unbound moisture exceeds bound moisture. When there are two methods, the solid is subjected to thermal drying at the same time:

- Move steam, primarily from the ambient atmosphere, to evaporate soil moisture.
- The imbalance between endogenic moisture and the subsequent evaporation by the application of energy to the solid surface.

The rate of drying can also be tracked by the continuation of both procedures. Concentration, convection or radiation and, in some situations, the combination of these factors may result in the transfer of energy as heat from the ambient atmosphere to a moist solid. The form and style of the different dryers vary according to the primary heat transfer mechanism. Heat is transported to the solid, moist surface and, in most situations, to the inner surface. Power is, however, produced in the inner solid by dielectric frying, RF or microwave drying and flows to the outer surfaces (Mukhopadhyay & Nateghi, 2017). The elimination of moisture from the surface of the substance by Method A depends on the environment, such as the temperature, humidity and air flow, the surface area and the pressure observed. The mobility of moisture inside the material in Method B is the product of the physical composition of the substance, the rate of diffusion and the moisture content of the sample. In addition to environmental factors, the method of drying agricultural crops during the drying period depends on:

- Exposure
- The Type and the Scale
- Initial Moisture Content
- Final moisture content
- Volume density
- The width of the screen
- Loop Transforming
- The temperature of the grain
- Temperature, humidity of the air in in contact with the grain
- Pre-treatment with industrial or prescription therapy
- The degree of air in agriculture

The drying speed can be influenced by both of these processes during the drying process, all of which can be carried out at the same time during the drying phase.

6.2.4 Early work on solar drying in India

Solar drying has altered the traditional method of sun-drying used in all agricultural crops. Initially, solar dryers were used for non-agricultural crops and were subsequently used for dry field products.

6.2.5 Solar Dryer

Sun dryers are designed to provide maximum heat under the requisite humidity, i.e. higher than ambient heat. The vapor pressure of the residual moisture in the material increases and the relative humidity of the air decreases in order to increase the ability of the air to maintain moisture. Air is pulled into the dryer by natural convection or by a fan. The moisture content of the material is heated and then gently refreshed when it reaches the collector. The substance of the wind and the sun is heated. The contents are directly heated. In order to preserve relative humidity, the temperature of the heater in the collector and the average humidity of the air prior to its admission into the collector will increase the humidity of the hot air compared to the cold air. Based on the initial humidity, the humidity and temperature at which it is heated.

Classification of solar dryer: Dryers are usually classified into two major working temperature groups between high temperature and low-temperature dryers. However, fossil fuel dryers (usually classic dryers) and solar energy dryers are commonly used. All high-temperature drying systems are primarily known as fossil fuels, whereas fossil fuels or solar energy systems are low-temperature dryers. In order to differentiate between various types of complex systems of solar dryers and different modes of operation, basic requirements must be changed (Ouedraogo et al., 2012). The following methods can be used to distinguish solar dryers: air flow system, air flow direction: ventilation, solar insulation proximity: thermal system area: system of air passage. Thanks to their heating systems and the use of solar radiation, solar dryers can also be classified. Normally, they can be classified into two major groups, i.e.

- Solar Sensitive Drying Machine (most of which are often termed hybrid solar dryer).
- Clean Solar Passive Unit (conventionally termed natural-circulation solar drying system).

The effectiveness of solar dryers depends heavily on environmental conditions. In certain situations, solar power supplies the heat required to absorb humidity and the energy used to drive fans. If the product is pre-treated, the temperature of the air determines the ability of the product to dry for a certain period of time. The drying time is shortened under sunny conditions and is also extended under extreme weather conditions. The disparity in drying ability during a downpour and rainy seasons must be taken into account when determining on annual drying outcomes.

A single energy source for small dryers, in which there is no risk that huge crops will be ruined by poor weather, is proposed for the use of solar energy. A dryer with a backup radiator is widely recommended for the use of large solar drainers for commercial use in bad weather conditions. Solar dryers must dry a wide range of fruits and vegetables. Owing to non-direct light (in house or cabinet-dryer) and a shorter drying time, up to a third of dry goods are excellent in consistency with traditional sun-drying.

Because a number of problems of constancy may be affected by the lack of a range of drying conditions and facilities, the drying process cannot be treated as an exclusion of moisture.

Any of the attractive qualities of dry goods are as follows:

- Low and stable amount of humidity
- Minimum broken and incomplete ratio of grain
- So, the lenient tendency to split
- Increased viability; increased viability;
- Weak number of molds
- Well, an essential diet
- Equal appearance and organoleptic features of consumers.

Even if technical innovation is needed due to economic conditions, work constraints or material shortages, it is incredibly unlikely for farmers to use the technology suggested. The State is providing financial and technological support for the production of solar drying companies in India due to its potential.

Advantages of a solar dryer

- Increase the drying time with high temperatures, air ventilation and humidity.
- Meat is used in the dryer and is also covered by pollen, beetles, birds and wildlife.

- The higher temperature kills insects and the faster rate of drying lowers the chance of spoilage of microorganisms.
- The higher drying rate also provides a higher feed capacity and a smaller drying area (approximately one-third).
- The dryers are waterproof and the food must not be moved during the season.
- Drying machines can be constructed from materials available locally and are reasonably cheap.

Limiting issues with a solar dryer

- It can only be used during the day when a reasonable amount of solar energy is available.
- Failure of skilled professional and maintenance staff.
- Little consistency relative to modern dryer versions.
- A backup heating system is needed for products that need continuous drying.

Design Consideration

- Air collectors: The solar collector of air is designed to heat the air by sunlight. Covers, absorbers, air passage and insulation are the core elements. The solar radiation produced by the cover insulates the absorber and then heats the air through the air. The typology of the complete air collector can be defined as follows on the basis of its substructures:
- Absorbers:
  - Non-permeable
  - Underflow
  - Decay of the ventilation fins
  - Pour over
  - Overflow and overflow
  - Glazed Matrix (Metal of cloth)
  - Unglazed, perforated aluminum
  - Collector for openings
  - Hybrid (air/water)
  - Hybrid (Photovoltaic/Air)
- Absorber coating:

  - Selective
  - Non-selective
- Glazing:
  - Single one, just
  - Double-Twin
  - Unglazed

Orientation of the Solar Absorber

The direction of the solar panel depends on the location of the service. It is commonly known to be located at an angle of latitude that goes through that area. It should be facing south if it is in the northern hemisphere, and if it is in the southern hemisphere, it should be facing north. It should be faced in the vertical orientation of the points in the equator.

Ventilation

The overall efficiency of the machine depends on the air flow volume. More air flow consumes more energy and less air flow creates less thermal power. The higher the rate of general circulation, the higher is the collector's productivity. The electrical energy of the fan increases with the mass air flow rate. The leakage effect increases with the air flow rate. The description alluded to above is based on different types of sun dryers:

Tent dryer: solar tents consist of a pulley frame covered in wood and are easily assembled and low-cost. Black acrylic has to be added to the wall facing the light. The material to be dried is on the top of the base. The drying process takes place at the same time as the open-air drying process. Dryers play a primary role in the control of dust, soil, moisture, wind and/or pests and are typically used in vegetables, fish, coffee or other items that are otherwise highly waste-related (Oughton et al., 2018). The tent dryers may be withdrawn and set up during non-operation. They've got the downside of being easily destroyed by strong winds.

Box dryer: A solar dryer box has commonly been used for small-scale food drying. It has a wooden box with a transparent lid attached to it. The inside is black and the mesh plate covers the dryer floor of the product. The air enters the chamber, leaving the vents at the top of the back wall from the holes in the front of the chamber.

Solar cabinet dryer: a large box of wood or metal is mounted inside a drying case on trays or shelves. The dryer shall be designated as a mixed or direct solar dryer, whether or not the chamber is translucent. The dryer is referred to as a diffuse type or an indirect solar dryer since the chamber is opaque. In the case of mixed dryers, the integral (direct) form characteristics and the (indirect) diffuse process of solar dryers are combined. The cumulative impact on the material to dry immediately from the solar radiation incident and the hot air provides the heat needed for the drying process. In a few cases, the air circulates through the air pipes from the drying chamber and from the drying trays containing the crop. The air is heated by a low-pressure sun collector. At the end of the chamber, moist air is then expelled by an air vent or a chimney. Proper and robust screening to minimize thermal losses (within economically justifiable limits). Aluminum or waterproof frames with cladding, whether paint or resin, are favored. Heat air flows through the tray stack until the material is dry. Second, once the hot air drops into the base tray, the tray will dry. The last chamber to dry is the roof of the chamber.

The system has advantages:

- The cost of work is very high. Then load and unload. The material is not necessarily susceptible to loss of color and vitamins. You should use a thermal storage unit.
- The inconvenience of this approach is as follows:
- A propensity to dry the bottom trays.
- During subsequent drying procedures, when most trays have been washed, fuel consumption is low.

Active solar cabinet drying systems: Active solar cabinet dryers are also known as hybrid solar dryers or forced convection dryers. Maximum air pressure in the dryer must be used during the drying period, regardless of atmospheric conditions, to retain temperature and humidity in wide areas. The volume of the airflow can be controlled and the mass profile is less constrained. Compared to natural convection dryers, the power and reliability of the dryers are also significantly improved. Well-designed solar dryers for forced convection are usually more efficient and controllable than natural airflow modes (Peacock et al., 2014). The use of forced convection would cut the drying time by three times and 50 percent less storage capacity. The six-fold higher collector fan-based dryers will also achieve the same quality as the natural convection dryer. Supplies can be powered by electricity or solar panels, if usable. Almost any type of natural convection dryer may also be powered by forced convection.

Back-up cabinet dryers: solar dryers normally heat the air using solar energy and improve performance on sunny days only. One of the key drawbacks of solar dryers is that they are not widely used in heating systems. This hinders the ability of commercial farmers to process crops, whether the weather is poor or overnight. The drying time can also be prolonged throughout the day if the sun is adequate. This not only decreases demand but also reduces performance. Efficiency may be processed on an ongoing basis for commercial farmers to meet market demand. Biomass, especially wood, is the most prevalent source of energy in rural areas of developing countries, as there is no overuse of local resources. Wood may be a source of green energy as new trees contribute to their use. Rapid, low-cost combustion equipment is also often needed to complement suitable solar technology such as cabinet drying systems. It's still currently pricey to burn.

Greenhouse dryer: the concept of a greenhouse dryer is designed to incorporate the role of the solar collector into the greenhouse structure. This solar dryer can be constructed from transparent materials such as glass, fiberglass, UV or polycarbonate panels, all of which are made on the roof and wall. Transparent materials are mounted on a strengthened steel frame or column with bolts and nuts and closed down to block the entrance of moist air or rainwater into a chamber other than the inlet opening. The system would use black surfaces to optimize the absorption of solar radiation. Entrance and exhaust ventilators are installed within the system to ensure that the dried air is similarly circulated. Greenhouse dryers can have greater control of the drying process when properly designed than dryer cabinets, which could be better suited for large-scale drying. A natural convection greenhouse-dryer was modeled by Ekechukwu, which included two parallel rows of galvanized iron-wire mesh drying platforms (long later). The fixed, inclined glass roof on the platform allowed the object to be exposed to sunlight. The dryer was coated, longitudinally aligned, with black inside walls to the north-south for better sunlight absorption. A crest cap, made up of a folded zinc tube, is placed over the roof. The air intake operated the shutters on the outside of the platforms (Perwez et al., 2015). A semi-cylindrical, transparent drying room with a vertically mounted cylindrical fireplace from one end is a modified version of a conventional, natural-type greenhouse solar dryer, while an air inlet door and a drying room are provided at the other end. The chimney can be over 3.0 m in height and 1.64 m in diameter. Constructed for variable height distribution. The drying chamber was an improved and extended version of the polytunnel greenhouse currently available.

The dryer works by contacting the material directly inside the dryer with solar energy. The material and the hanging black absorption curtain, which heats the indoor air in response, absorbs and heats solar radiation within the chimney. When warm air rises and the fireplace sinks outside the dryer, fresh air fills the other side of the dryer. Apart from the clear advantages of passive solar energy dryers over active types, they also offer low cost and flexibility in the construction of the site as well as on-site operating systems for the natural ventilation of the ventilated solar energy dryer compared to other passive solar energy dryers (for installations in rural farm locations in developing countries). Its main drawback is that heavy wind power interruptions are vulnerable.

6.3 DIRECT AND INDIRECT SOLAR DRYING

According to FAO estimates, more than 300 000 tons of mangoes are produced per season (mainly in Ivory Coast, Burkina Faso, Mali, Senegal and Ghana). However, a significant proportion (about 150 000 tons) of this production was lost during and after harvest, compared to comparatively small volumes exported mainly to the EU. From 2003 to 2006, these exports accounted for about 6% of total production and cost some countries more than USD 60 million. Mango trading would also offer huge financial gains to these poor countries if losses were to be minimized after processing. The drying process is used more and more to minimize losses after harvesting, to keep some of the fruit supply until it is sold. Absolute and indirect solar drying is also increasingly accepted as harvest time (extreme heat, low relative humidity) and other high energy source prices are favorable (electricity, gas). Any of these methods of solar drying have been independently checked, but no distinction has been made between their effectiveness and characteristics.

In our current study, we have decided to address this comparative review. Recent experiments on the solar drying of West African mangoes have been conducted: a thin-layer, indirect solar drying model has been reported by Disso with experimental confirmation of mango slices. Rankins, Sather and Spicer performed sun-drying experiments with Senegal using a greenhouse-type solar dryer. These writers have shown that solar-drying mango can be a means of preserving an essential supply of vitamin A to the French-speaking western population (Rosenberg et al., 2013). In an experimental study of a typical convective heat wave solar dryer, Gbaha studied direct solar drying kinetics for plantain bananas, sweet bananas, pineapple and mango. Koua conducted statistical modeling of the thin layer of banana, mango and cassava, with seven computational semi-empirical simulations, and studied the drying parameters of these products. Touré and Kibangu-Nkembo have reported a comparative study of direct solar cassava, banana and mango drying. The original moisture content was compared to the maximum difference between the temperature of the material and the temperature of the ambient air. The objective of the current study is to compare direct and indirect solar dry parameters produced in West Africa for key mango species (Amelie and Brooks). In both drying methods, drying characteristics were evaluated and compared, including drying curves, drying rates, diffusiveness of the water and drying consistency. Finally, the color measurements and the final water activity value determine the accuracy of the dry product produced during each drying process.

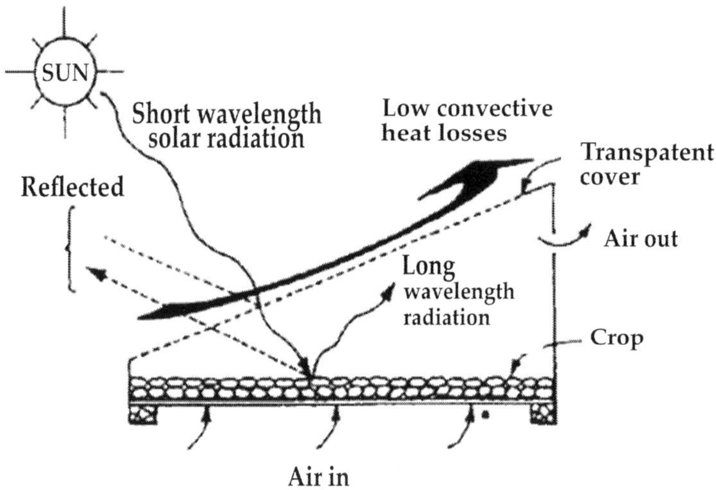

Figure 6.2: Working principle of direct solar drying.

Source: https://www.researchgate.net/figure/Working-principle-of-direct-solar-drying_fig3_315542121

6.3.1 Materials and Methods

Experimental Dryers

Direct solar dryer: A 1.20 m metal frame of four rectangular rays with a scale of 0.39 m x 1.02 m was made of a direct solar dryer. The trays were designed with a wooden frame, covered by a nylon mesh, to allow the passage of air through the dry bed. The overall capacity of each drying tray was 2 kg. The sides of the dryer have also been covered with nylon mesh to avoid friction between the drying item and the outside. The last tray in the upper section of the dryer is vulnerable to sunlight. Solar radiation is often used for this type of dryer as heat is transferred to evaporated product water. The ambient air that flows into the drying space produces water vapor. The effectiveness of such a dryer depends on the ambient and solar radiation at the drying point.

Indirect solar dryer: The solar collector (useful for the generation of thermal energy) and the dryer are installed on the indirect solar dryer. The solar collector had a mixture of absorbers caused by the joining of a rectangular duct with a corrugated iron absorber and a porous aluminum mesh absorber. A part of the solar radiation emitted from the surface of the array is transformed into heat. The air flowing through the porous absorber absorbs a portion of its energy through a convective heat exchange. The drying process is used to dry the mango slices using hot air removed. This drying system is a parallel wooden enclosure. The top is enclosed by a cone-shaped roof with a PVC chimney apex. This boiler supports a low-power extractor (12W) and an airflow extractor of 64, 8 m3/h at its middle (Raymond et al., 2019). The rectangular opening for the hot air inlet is made in the lower part of the drying unit. One side is designed for loading and unloading mango pieces from the dryer by means of a swing door. Two superimposed 5 cm glass-wool coats and a 7 cm rasp distinguish the lower lip. The whole device is covered by a metal frame, 80 cm above the ground. The products are set in four rectangular tables inside the dryer. Each tray is made of wooden frames with a fixed nylon net that encourages air movement. The trays are 20 cm apart. The top tray is 40 cm from the roof base and the bottom tray is about 20 cm from the insulation sheet. Up to 2 kg of space is provided in each of these trays.

Mango Samples

The mangoes used in this study are Amelie and Brooks. These two big mangoes have been grown from April to June as well as from the end of June to August. They are about 50% of the amount of mango sold in West Africa. Strong quality fruit was obtained from the local fruit market, washed and peeled for drying in Ouagadougou (Burkina Faso). The flesh of each fruit was extracted by means of a stainless steel pin and divided into 8 mm thick samples. The slices were then loosely scattered over the solar drying trays.

Experimental Procedure

The University of Ouagadougou (Ouagadougou, Burkina Faso) has repeated direct and indirect solar drying operations at latitude 12°20-12°26 N and longitude 1°28-10036 W at the same time, in April, May and June 2006 and 2008. The study produces direct solar drying results for Amelie on 21, 22, 23 and 24 May 2008, as well as for Brooks on 26, 27, 28 and 29 May 2008. Indirect solar drying data are also available for Amelie on 21, 22 and 23 May 2006 and for Brooks on 226, 27 and 28 May 2006. For each drying process, an OHAUS balance weight of 10 kg was carried out at a maximum capacity of 0, 1 g. The CD11 model indicator linked to that balance allowed a direct reading of the drying rack mass. Each dryer had four tables numbered 1 to 4 and waved from bottom to top before each drying process had started. The mango slices of the four trays, which occupy an estimated 60 percent of the overall area, are evenly scattered for drying. At the beginning of the drying process, each tray, containing between 350 g and 800 g of mango-flesh, housed direct and indirect solar dryers (Simoes et al., 2014). During the drying process, the trays were withdrawn regularly and weighed at 1 hour at the beginning of the process to 2 hours at the end of the procedure. The properties of the drying air were also calculated in each event. The system was aided by the K-type thermocouples with TESTPOINT temperature control software (version 3.4) and the relative humidity was measured with Bio block thermo hygrometric accuracy of ±3 percent at indoor temperatures, ambient air and indirect solar drying.

6.4 HEAT-ASSISTED DRYING

Drying is a major tool in various food sectors and in many agricultural countries with huge volumes of dried food products planned to increase shelf life, decrease manufacturing costs, reduce shipping weights, boost attractiveness, encapsulate original flavorings, and retain nutritional value. The primary aim of drying is to remove water from food so that food cannot grow and so that bacteria, yeast and mold are eliminated. Economic understanding, environmental concerns and facets of quality control are the main three-fold subject of research on drying processes in the food industry. In certain countries around the world, the high demand for dehydrated fruit and vegetables is especially high.

Clean Indian exports to the rest of the country are stored. 424 (USD Million 94, 07) crores increased from 364 in 2005-06 to 11 in 2006-07. Russia, France, the United States, Germany and Spain are the main importers of dried and processed Indian vegetables. Good long-term storage of fruit and vegetables is also necessary in order to achieve high economic prices. In order to do this, these products must be dried at a low moisture level (Shove& Walker, 2014).

However, drying prevents but does not inactivate, the development of enzymes (naturally occurring compounds that make food mature). Such drying methods also restrict the amount of nutrients, such as vitamin C loss, which does not help to improve color and appearance. Air drying is an example of this, however, causes significant harm, such as enhanced taste, color and nutritional value of meats, reduced density and potential absorption of water, and solutes are transferred from the inside of the drying product to the surface due to lengthy drying times and high temperatures. Drying should be in control mode in order to prevent any such inconsistencies.

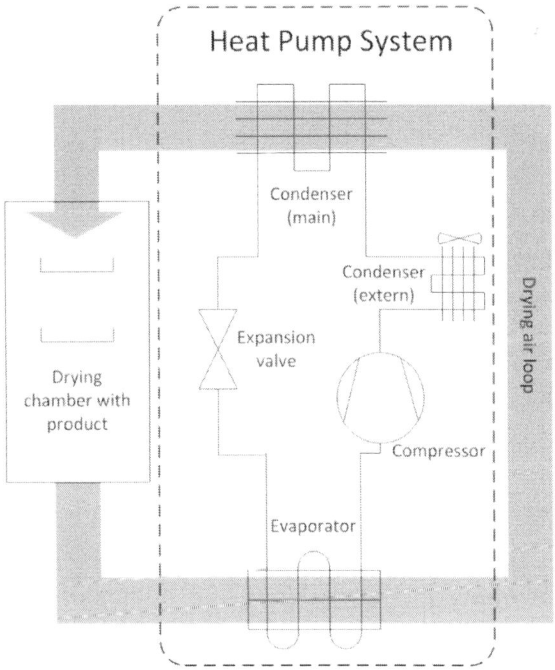

Figure 6.3: Principal layout for heat pump assisted drying.

Source: https://www.researchgate.net/figure/Principal-layout-for-heat-pump-assisted-drying-for-convective-air-drying-in-closed-loop_fig1_314096009

Yet courage is the word "goodness" for all of us, as we embrace and integrate it into our everyday lives. If people and every single living human are in a healthy place, they still feel extremely comfortable. When we conducted an experiment/research in a controlled atmosphere/process, the result was excellent as an unregulated environment/process. The result would not only be economic, cost-effective and time-saving, but also high-quality, if we were to think about regulated technology. We will also save money and time by using this technology, as is needed today. One of the advances in temperature and humidity is the heat pump-assisted drying process. Other common approaches in food industries include spray sedatives, freezers, vacuum dryers, tray sediments, rotary sediments and fluid bed sediments, etc.

Drying is a former, traditional and renowned food preservation unit or activity. However, demand for fruit and vegetable dries has declined because of the use of energy and time, i.e. higher moisture levels in food products are more likely to die. Since it can be retrieved and operated at the lowest rate, heat pump-assisted drying is energy efficient. However, the cost-effectiveness of the heat pump dryers (HPD) has not been apparent to date due to the use of high energy (electricity) and the question is now being asked as to whether the dryers are more feasible and efficient and how to understand this, etc. Despite the designation, a wide variety of fields, including conventional trays, rotary trays, blaze, spray, fluid liner, vacuum, freezer and batch dryers, are suitable for further development. Mujumdar has outlined a number of areas and aspects that require further research and development (Sørensen, 2008).

This section also illustrates more clearly how heat pump-assisted drying can be done to fulfill more of our existing needs. The fastest growing method of agricultural goods. In this post, we discussed why heat pump drying should be financed. Owing to the low energy consumption of drying devices and methods, this can be reduced. The operating costs for hot convective dryers and direct dryers are lower for prasertsan and saensaby (HPD). HPD drying has been shown. Reay and Macmichael, McMullen, and Morgan and Heap have been used for fast heat pump applications and methods.

6.5 HEAT PUMP ASSISTED DRYING

The heat pump-assisted drying process is a combination or mixed drying system. Hybrid or hybrid drying systems involve the introduction of various heat transfer modes, two or three phases of the same or different dryer form.

With the addition of heat pump-assisted drying, heat pump-assisted drying is used in drying methods that can be produced in several different ways and there are also some of the most promising techniques, heat pump drying and heat pump-assisted drying that have been used so far in the food industry. Intense research of these processes is determined by the drying efficiency of both electricity, process time and time. The amount of heat absorbed by a wet substance in a dryer varies from 5% to 10% of the energy supplied. With the aid of a heat pump for drying, 95-90 percent of this energy can be saved, leading to immense energy savings. Vázquez suggested that heat pump dryers can be used as an alternative system for drying fruit and vegetables with lower energy usage, relative humidity and temperature values. The low temperature and improved heat pump power provided by the dryer showed many benefits, according to Best. The drying assisted heat pump is therefore an ideal alternative to traditional drying systems (Silvente et al., 2015). Tai has carried out a test of the heat pump dehumidification unit R-114 and it has been shown that the heat pump assisted drying is more successful and that a number of scientific tests of the heat pump assisted dryers can also be found in the literature.

Cartwright identified the modeling of the heat pump-assisted dryer for improved efficiency, and Baines and Carrington conducted a simulation analysis in which various configurations and effects of basic design improvements were studied. Two components, namely the heat pump and the actual drying material, are treated separately during the heat pump drying process. In the literature given by Reay and Reay and MacMichael, the basic definition, principles and mechanism of heat-pumping assisted dehumidification can be found. In conjunction with other drying methods such as spray drying, atmospheric freeze-drying and solar drying, heat pump drying can also be carried out.

6.5.1 Principle and Mechanism

The basic theory of hybrid (heat-pump) drying is the same as that of a heat pump (pure) thermal pump and can be referred to as a reverse operating heat generator. The heat pump uses the cooling principle that cools and condenses the flow of air. The air is kept clear and the resulting evaporation heat is recirculated by collecting water vapor. Heat pumps, says Gac, are thermodynamic devices designed to heat at the appropriate temperature by using heat obtained from the source at a lower temperature. The heat pump is also an apparatus for the use of low-temperature energy sources in and on the surface of ambient air, lake or river water. Heat pumps can also be run with 'waste heat' from commercial and industrial activities for some thermal processes, thereby rising to the necessary temperature level and because the heat pump has both heating and cooling capacity. It absorbs heat from one medium at a lower temperature and then moves it to another medium at a higher temperature (a high-level sink).

The heat pump dryer consists of the normal drying chamber of the air circulation system and its regular parts. The dry air is removed and reheated by the evaporator using the heat pump condenser (cooling component of the cooling cycle). The condensation temperature shall decide the average drying temperature of the coolant to be used.

Additional developments have been made in the heat pump drying of loop thermosyphons, which is a form of heat flow. Parts of the evaporator and condenser are linked to the thermo-symphonic circle, and all pieces are bound to the string. Due to its severity, the thermosyphon is positioned around the Phaphuangewittayacul heat pump dryer evaporators when the work fluid is diluted and the air is pre-cooed.

The thermal properties of the HPD and its efficiency can be calculated by the sum of thermal energy from the evaporators, the MER (humidity extraction rate, i.e. kilograms of humidity extracted every hour) and the SMER (humidity extraction rate per kilowatt hour) indicating the accuracy of the energy used (coefficient of performance, i.e. the ratio of heat rejected by the condenser to the work used by the compressor).

In conclusion, agriculture is the driver of economic development in most African countries. But when it comes to money and family eating, growing crops are only part of the story. Equally important is what happens during food production and harvesting. Any profit means the addition of value, the creation of crops or animals into niche-appropriate products. Drying is one way to make agricultural goods more valuable. Goods are available with the right services and expertise at a fair price both domestically and abroad. There's a trick on drying well, though. Successful, extensive and important to our diet and our animals are the limited vitamins we use. Due to surpluses during harvest, a substantial portion of the food provided by farmers on the market will have been wasted (Silvente et al., 2012). Drying can also be a very effective way to store and increase food and nutrient demand and potential benefits. Drying is not without the prospect of fruit, vegetables, meat and fish.

Dry products should be eaten fresh relative to other food products. And they're a food-borne infection vector because they're raw. Hygiene is also very critical in this area-handling, washing, sorting and packaging. This kit addresses some of the challenges involved with the storage and drying of field products for health and hygiene. It is important to use boiling water to wash food before cutting and frying. When germs spread easily, you should ensure that separate areas are used for the washing and cooking of raw materials. This resource pack can be drained for almost every sort of vegetable or fruit. Both meat and fish taste healthy and dry. We can't try to cover all the different kinds of products that this pack would dry up! However, we should discuss different drying methods for export both locally and abroad. We will discuss how to build some of the most powerful drying methods. For example, sun-drying may be an important way of recycling, because such crops are sold on the market at harvest time, so that food is scarce and can be stored and taken. Listening to a solar drying interview, you will understand that measures are being made to make it easier for your listeners to make the most of the dry field products. Investment is required to develop solar drying technology so that you can make more of it from your dried goods while the structure is in operation. Another solution that we should look at is the pretreatment of food by preserving and drying animal feed. In any case, interviews are intended to educate the audience about the most cost-effective way to dry and sell small-scale farms in order to make more profit.

Chapter

7

Cooling and Cold Storage Energy for Agricultural Production

CONTENTS

7.1 INTRODUCTION

A number of food items, accounting for about 30%, that is consumed by individuals in developing countries is perishable. As such, in the post-harvest handling chain, cooling is considered as one of the most important steps. The process by which one reduces the temperature of the output can help reduce the respiration rate and further extend the shelf life which then leads to the protection of the quality. With all this, the loss in terms of the volume of output will be reduced. The science behind cooling is that there is a reduction in the rates of water loss and decay which prevents the output from withering or even decay with its moisture content, especially when the produce is covered. The step which is taken to cool the produce immediately after the produce has been harvested is what is known as pre-cooling. The name given is because of the fact that, the produce is stored in a cool place before they are transferred to cold storage or even put in a refrigerated truck or in other cases marine containers for the purpose of transportation. The use of the cold storage facilities is considered to be very crucial especially when one needs to minimize post-harvest losses. However, in all instances, cold storage facilities cannot be used independently as losses occur in every step of the post-harvest cycle (Utama et al., 2014).

As such, cold storage is one component that has been integrated into the cold storage network from the point in which the produce is harvested to the point in which the consumers purchase the produce. The different stages in the post-harvest cycle include the harvesting itself, followed by precooling, there is then the transportation and packaging, then cold storage then finally the important part which is the display in the market so that the consumer can be able to purchase the produce of their own choice.

Figure 7.1: The majority of horticultural production are perishable and needs to be stored in cool temperatures in order to ensure that their shelf life is prolonged.

Source: Study.com

On these different stages, the temperature is not the only thing that is considered even though very crucial, there is also need for proper handling of the products which is involved in the process of cleaning, adequate packing and also sorting of the same before packing. This section however considers the cooling processes and cold storage of the agricultural produce to prevent post-harvest losses. As a matter of fact, the shelf life of produce in terms of post-harvest depends entirely on whether the fresh produce from the farm is kept cool. Ideally, when the temperature around any produce increases, there will be a significant increase in the deterioration. When this happens, there will be higher post-harvest losses which will be a loss for the producer, the suppliers and also the consumers using the product. Furthermore, the deterioration of the product caused by high respiration rate, mechanical injuries, pathological breakdown, ethylene production, water stress and physiological disorders leads not only to the decay of the product but also changes in color, flavor and texture of the products and most important the loss of the nutritive value of the produce.

Figure 7.2: Plants are living organisms and once harvested, they are still living and thus need to be stored properly after they have been harvested to prevent decay.

Source: Britannica

Some of the spoilage reasons such as molding and pest infection among others thrive in warm conditions, which normally occur when the produce is not well stored in cool temperatures. Most of the factors that lead to the deterioration of the harvested produced are commonly addressed by the process of cooling and cold storage. Some of the physiological changes such as ethylene production, enzymatic process and respiration can be highly lowered by reducing the temperature surrounding the produce. Furthermore, the reduction in temperatures can also reduce the growth of microorganisms such as fungi and bacteria. Plants are living organisms and most of the produce after they are harvested are considered to be still alive and thus the metabolism process is still taking place and also the various activities of the microorganisms are also continuing that easily leads to the deterioration of the product. In order to reduce the rates of respiration in terms of the produce, one needs to delay the ripening of the produce in order to reduce water loss and this can only be done by reducing the temperatures surrounding the produce. Having adequate and efficient storage facilities can to some levels control some of the factors that negatively influence the produce hence improving the product quality, the value of the product and for certain the shelf life of the produce. In order to maximize the quality and the shelf life of the produce, the factors involved should ensure that the produce is pre-cooled after harvest, transported then placed in cold storage (Uchiyama, 2002). Depending on the type of produce, optimum storage temperatures for the particular produce should be considered. Ideally, the storage temperatures that are considered to be optimal are considered to be as low as possible but should still be above the chilling temperatures that are considered to be sensitive for the particular crop in question.

7.2 IMPORTANCE OF COOLING AND COLD STORAGE

In the entire post-harvest cycle, cooling is considered to be the most important step and also is considered to be the most energy-intensive segment in the food industry. This is so because the cooling and cold storage use a lot of energy or rather huge amounts of power is used to make sure that the produce is kept cool. Take for instance one appliance that is used in our homes to store products to ensure that the shelf life is prolonged. The fridge of course is one appliance that is found at the homes of members of the public with the aim of keeping perishable products fresh. The fridge uses a lot of energy as witnessed by the people owning the same by the power bills they pay to the electric companies. However, the mechanism behind the workings of the refrigerator is quite complicated. This is so because, perishable products need significant levels of refrigeration and heating in order to slow down the spoilage process and further maintain the freshness of the fruit post-harvest, its flavor and also the levels of nutrients. When cooling of the fresh fruit is done before it is processed, the main aim is to remove the heat from the field in time to ensure that there is the preventing of decay of the particular product. Furthermore, with cooling, there will be the retention of the moisture content, the vitamins on the fruit, the sugars and the starches present. When it comes to the cooling of fresh vegetables including broccoli, asparagus and even cauliflower, the main aim is to maintain the color of the vegetables and also their texture, that is, make sure that when they are sold, they are firm.

It is clear that most people may have heard about the cold rooms in the movies or have been told of the cold rooms in restaurants. However, the majority of the people do not know what exactly is entailed in the cold rooms. First, it should be understood that the cold rooms are the rooms in which lower temperatures are maintained at all times whereby the temperatures are able to be maintained with the help of what act as precision instruments and also other machines fitted on the walls of the room (van Ruijven et al., 2019). In the current world, there is always an evolvement of new technologies and the fact that cold rooms are always being built and continuously being improved cannot be regarded to be surprising. The work on cold rooms is continuous and is also given much priority because of the fact that they are responsible for ensuring that the food meets the demand and supply game when considering the food products that are perishable.

It is with no doubt that there are a number of facilities that use the cold rooms but the food industry makes the top of the list. The aim of building these facilities is especially to store products that are perishable for longer periods of time. With the increase in the world's population, there is an increase in demand for food products, especially those that are perishable, which are fruits and vegetables. So as to ensure that these products do not get spoiled, there is a need for cold storage facilities (Wang et al., 2011). One important function of the cold rooms is that they tend to help prolong the shelf life of different fruits and vegetables and are also considered to be very effective in reducing the amount of waste and also helps lengthen the time frame of the food products in the market.

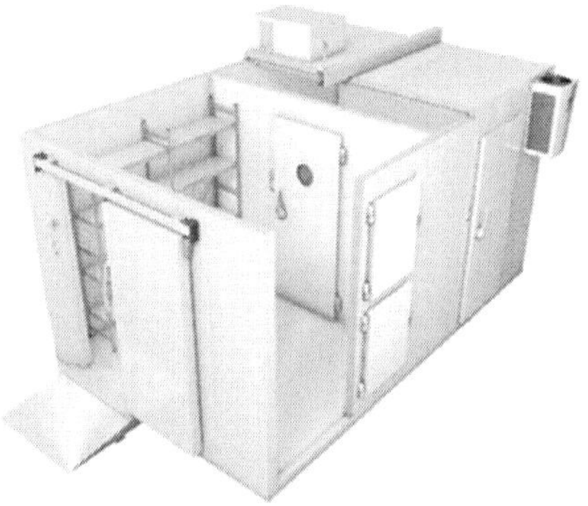

Figure 7.3: The cold rooms are normally made for food processors but should be designed according to the client's need and purposes.

Source: Kinsman ltd

When one is constructing the cold room facilities, they are normally made to meet the client's needs and purposes for the cold room. Care is taken and every little detail is taken into account when dealing with the building of the storage facilities. These projects are normally meant for the food and drink industry in order to maintain them. They are made for the following services, hygienic floor finishes, chiller and freezer cold stores, food and drink manufacturing facilities, drainage installation, service and preventative maintenance among many others. As such, the processors of perishable food products are asked to consider the designers of their cold rooms before deciding to hire one.

The above-stated explanation shows that the cooling of fresh vegetables and fruits helps maintain the moisture content of the fruits, the nutritional value of the fruits, the quality and most importantly, maintain the physical properties for longer periods (Waibel et al., 2019). The process of cooling and cold storage of fruits and vegetables in the refrigeration systems, especially for those involved in the processing of the products can prove to be very expensive. This is so because they tend to operate at heavier loads, especially during the summer when the amounts of fruits harvested are quite large. At these times, the electrical costs tend to be high as the refrigerators are required to operate at heaviest loads given that the outside temperatures are normally very high. There are various ways in which cooling of the perishable products can be done which will be explained further.

7.3 MANUAL/ LOW-TECH TECHNOLOGIES

Figure 7.4: Assembly points need to be tented in order to maintain the humidity and thus cool the produce.

Source: The Postharvest Education Foundation

These involve the low-cost methods of cooling perishable food products. There are some manual methods that can be used in maintaining the shelf life or rather cooling the perishable commodities but are not considered to be very effective. One of the methods is shading. Shading involves a process whereby the fresh produce from the farm is covered to protect them from direct sunlight in order to prevent it from gaining heat. When it comes to the working areas, the farmers are recommended to have cloth or a roofing tent to provide the shade for the produce immediately after they have harvested the produce. The assembly points also require to be tented while they wait for cold storage of the products. The roof used should be deep and overhanging for at least one meter in order to prevent the light or rather the heat from passing through the windows and the doors. The roof should either be light-colored or rather one with a material having reflective properties as they have the ability to reduce the room temperatures or rather the temperatures under the shelter to about 20^{0}C.

Another method that is considered to be manual and one can only spend lower amounts of cash to cool the perishable fruits and vegetables is the use of cold water. The water is gotten from the deep wells or the mountain streams. In most parts of the world, one would observe that the water temperatures are normally quite lower compared to the air temperatures. The temperatures of water from deep wells and mountain streams are normally measured in terms of the annual temperature of the air in the particular area in which the water is found. In the storage facility or in other cases environment, water can be used for the purposes of hydro-cooling whereby the water can be used as a spray or a mist in order to maintain the relatively high humidity of the space. The only disadvantage of using well water is the fact that there is often contamination in a way and it is not advisable to use them on either fruits or vegetables.

The third method that can be used to cool vegetables is the use of passive evaporative cooling. Passive evaporative cooling of a cooling room is whereby one wets the walls of the cooling room or one can decide to use porous materials on a single end of a structure. Such porous materials may include the use of straw pads that are wet by water. This method can provide active cooling in the storage facility or rather in the cooling area by water evaporation. The mechanism behind it is the fact that the hot air is pulled through a wet pad or the wall fitted by the porous material by use of fans that are considered to be low-speed ventilation. The consideration is that this method is very effective in areas that have dry climates but may prove to be less efficient when the humidity is high. This method is used in large greenhouses.

7.4 FOSSIL FUEL-POWERED AND ELECTRIC TECHNOLOGIES

Room cooling. Room cooling is considered to be a simple and cheaper method but may prove to be very slow when used. It is normally used to reduce the temperature of the produce before it are transferred to cold storage. The cold room is whereby the produce is placed in which the produce are then allowed to slowly cool down. In most cases, room cooling is not recommended for produce that are highly perishable as this method is only effective after 24-48 hours after they have been placed in the room. During the initial cooling in the room, the fans are required to provide 90 CFM/MT. once the target temperatures have been achieved, the fan speed can be reduced in order to ensure that there will be air flow at about 18-25 CFM/

MT. When using room cooling the stack of the produce should be narrow. The installation of the fans should be such that cold air can have the ability to move around the room. The air that will be circulating in the room will be passing through the spaces. In this case, it shows that cooling in the center of the stacks is normally through conduction. In this case, there is a need for staff to check in order to ensure that the produce is being cooled as desired by the fruits or vegetables. In the start, one should be able to rearrange the stacks of produce while measuring the rate at which the produce is cooling until the point where one finds the right pattern for the particular cold room.

Ice cooling. In places or rather the locations where electricity and mechanical refrigeration is not available especially in an ice making and distribution plant, the systems allow for cooling in such locations. In the USA, the basis for long-distance perishable business was developed based on the ice business and has since then been very successful. The use of ice as a cooling agent is however very inefficient since only about half of the cooling effect is actually used in the cooling of the produce. The major portion of the ice is normally lost due to its interaction with warm air as it exchanges with the warm environment. Consequently, there can also be a loss of ice especially when it is transported from the central refrigeration plant to the facility in which the cooling of the produce normally takes place. Unless the manufacturer or rather the agent takes an effort to use packages in which ice can be placed in a well-insulated environment, such as what is known as an ice chest-style container, he/she might find him/herself losing at least 50% of the ice before even he/she gets to use the ice.

Figure 7.5: Insulated box cooling of perishable products to avoid melting of ice.

Source: Liviri.com

This shows that this method can result in a lot of losses and wastes a lot of time. This shows that this method is less inefficient when comparing it to other methods. An Ice machine is an energy-driven system and therefore ice making can be considered as very expensive. The Crytec's Bubble Slurry is considered to be one of the most efficient icemakers in the world, which has a capacity of producing 146 kg of pure ice in a single hour and produces about 19 kg per kWh. To cool about 1 MT of fresh produce at about 28^{0}C, 66 kWh is required, which is about 330 kg of ice. At this point, making ice of that sort for this purpose is considered expensive and thus prohibited.

Evaporative forced air-cooling. This is another way used in the cooling of fresh produce. This method involves the use of a wet pad and an electric fan. The components move the cool air through the containers of the produce in order to speed the cooling process. The temperature of the produce in this case is reduced by the use of evaporative cooling to a few degrees, which is the temperature at which the moisture content begins to form. Against a wide range of static pressure, the fan should have the ability to provide airflow of about 1L/se/kg. doubling the air flow is considered not to be a problem to the produce but at that point, the cost will rise at inconsiderable levels as the fan will be required to provide 4 times greater horsepower in order to have the ability to accomplish the same level of work.

Forced air pre-cooling inside a cold room. Another method uses fossil fuels to cool perishable produce. This method is generally known as FA cooling which tends to speed up the cooling of a certain batch of produce in a stack when inside a cold room for more than two days to a period of eight hours. If there is a cold room available with adequate capacity for the produce, if an individual decides to add a portable tunnel used for portable forced air, which can have the ability to cool 4 pallets at a time (Züttel et al., 2010). This will increase the fan's power in terms of use by only 800 to 1500 watts per hour. In a cold room that has 5tons of refrigeration 3MT of perishable products can be cooled from an initial temperature of 27^{0}C to a target temperature of 2^{0}C in about 6-8 hours. In order to accommodate airflow without excessive pressure drop, the area of the vents o the sides of produce containers should be at least 5% of the container surface area. The fans for the FA coolers tend to operate within the range of 0.5 -2.0 L/kg/sec. In order to speed the cooling process, one can do so by doubling the air flow rate but one should consider the fact that the cost associated with the energy used will rise since the fan will need to use more than 6 times the power that it normally uses. In addition, when the airflow is doubled, one should consider using a fan that is larger. Most industries tend to use centrifugal fans that have forwarded blades for the purposes of cooling. These industrial propeller fans are normally more suited for use with air pressure that is low.

Hydro-cooler. This method also uses fossil fuels to cool perishable produce. The water used for the purposes of hydro-cooling needs to be kept cold by the use of ice or mechanical refrigeration. Water is considered a much better conductor of heat when compared to air, which is a poor one. As such, the hydro-coolers are considered as fast coolers of the produce than the forced air coolers. When the hydro-coolers are designed in form of showers, the produce such as cherries that have smaller diameters can be able to cool for a period of less than 10 minutes. When it comes to produce that have larger diameters such as watermelon, take periods of about 45-60 minutes.

Figure 7.6: Modern hydro-coolers that use a conveyer belt to push the products from one end to another during cooling.

Source: TRJ Refrigeration

Hydro-coolers that are made in form of batch-styles tend to hold either one or more pallets of produce whereby cold water is showered over the top of containers that are stacked, which allows the water to filter down through the containers while in contact with the produce. The water removes the heat from the produce since it passes through the load. In a batch hydro-cooler; the door is normally closed during the cooling process in order to reduce the heat infiltration and water losses (Weiss et al., 2010). With the case of immersion hydro-coolers, which are rectangular, are shallow and large with tanks that hold the chilled water. The mechanism behind it is that the crates of produce are placed on one end of the tank and can be moved by the use of hand and by use of a conveyer belt to the other end of the tank where they can then be removed. The mechanical refrigerators or even the crushed ice is normally used to keep the water cold and a pump that ensures that the water is always in motion. Most produce are non-floating and so in most cases, the produce items tend to stay submerged in the water. The time that the produce stays in water tends to varies from time to time depending on the initial conditions and the desired temperatures at the end. The shower coolers tend to take a shorter time compared to the immersion type hydro coolers since the water moves past the produce at lower speeds. However, if the water is agitated, the cooling speed can be improved.

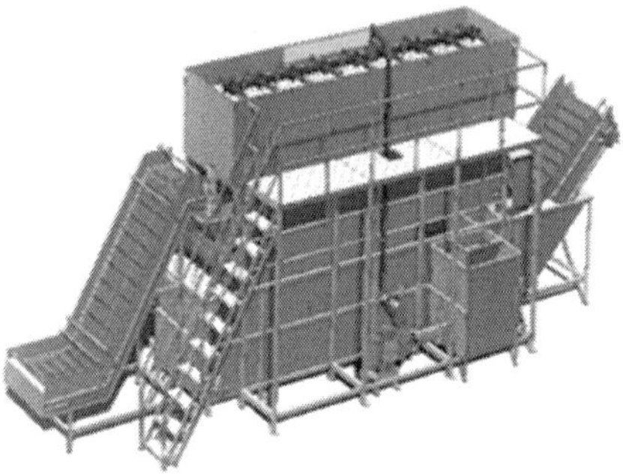

Figure 7.7: A modern vegetable hydro-cooling system.

Source: Novatech Engineering

The use of hydro coolers ensures that there will be a faster and uniform cooling of some of the commodities. The packaging material that is used for a certain product should be one in which the water can be able to penetrate or can wet in order to ensure that there is cooling and that the chlorine meant to sanitize the water reaches the fruit to ensure that the fruits are free from germs during the cooling process. The use of a packaging material during cooling is required in order to ensure that the fruit or vegetable is not damaged in the process of water beating. The hydro coolers that have tanks that allow the produce to be submerged are considered to be the best form as one can shower the produce with iced water and can also soak in the same water. In addition, since the shower is constantly pouring the water into the tank then, the water will always remain to be cool. This type of hydro-cooler can be constructed in a way that it can hold an entire pallet-load of produce.

7.5 COLD AND COLD STORAGE

In tropical and subtropical climates, cold storages tend to demand high amounts of energy. However, the total cost of the cold storage, that is, summing it up together with the energy demand are normally offset by the fact that there is a cost of saving since there are lesser product losses and also the fact that the produce is of better quality. As such, the storage space should be such that there is efficient insulation of the roof and the walls in order to reduce the amount of power that is needed. It should however be noted that different agricultural produce has different temperature requirements, especially depending on their biological origins.

7.5.1 Fossil Fuel Powered Technologies

Many people including main stakeholders in agriculture have not realized that enormous amounts of energy are required in the farming systems for effective production. As a result of farmers shifting from subsistence, mixed farming to commercial farming, most agricultural products require energy –intensive farming to cultivate optimally. Past studies have reported that fossil fuels have been engaged in large amounts to power heavy farming machinery, to process foods, to refrigerate foods during transportation, to produce packaging materials, and to manufacture and transport chemical inputs such as fertilizers and pesticides. In fact, fertilizers containing nitrogen have especially high amounts of fossil fuel where production and distribution of about one lb. of nitrogen releases an average of 3.7 Ibs of carbon dioxide into the atmosphere. These fertilizers have had a tremendous effect on agricultural yields, and a 2005 study of several types of crops found that the average yield attributed to fertilizer ranged from 40 to 60 percent in the USA and England and was reported to be higher in the tropics. Additionally, past records have shown that average U.S farms use a total of about 3 calories of fossil-powered energy to produce each calorie of food energy.

Figure 7.8: Fossil Fuels are Dead, Long Live Fossil Fuels.

Source: Energy Institute Blog-WordPress.com

However, other people have advocated for limiting or abandoning fossil fuels by a movement called "keep it in the ground" with the allegation that our expanding use of fossil fuels is fundamentally damaging the environment. The agricultural improvements over the past two centuries, which have allowed us to feed the rapidly growing populations while minimizing our footprint, show that this claim is indeed patently false. Besides, abandoning the use of fossil fuel technologies would more than double the amounts of cropland required to maintain the current levels of food. The fossil fuel technologies in agriculture have proved beyond a reasonable doubt that they are not only beneficial to humanity in improving access to affordable food, but also reduce our footprint on the land and helped in preserving nature.

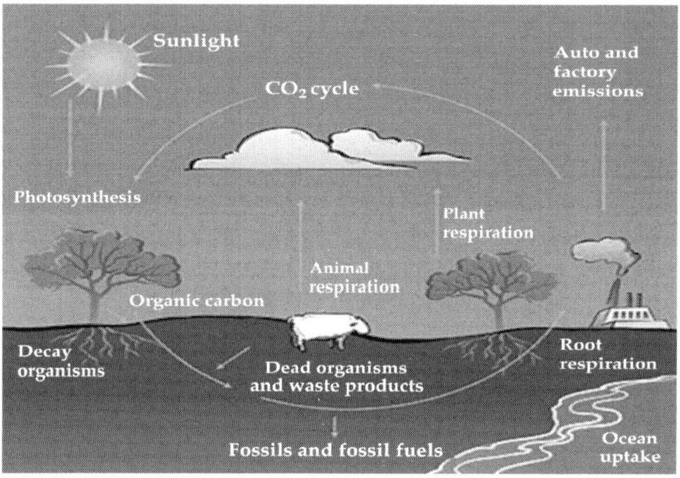

Figure 7.9: Fossil fuels in the ecosystem.

Source: Lenntech

Development in refrigeration that has been made possible by energy from fossil fuels has conquered the seasons and made fresh fruits and vegetables available to mankind even in the coldest winters and hottest summers. Moreover, they have contributed generously to improving the reliability of the food supply. Numerous synthetic materials from fossil fuels have been generated including plastic, asphalt, synthetic fibers, etc. which have assisted in storing foods and other products. Interestingly in almost all developed world, where these innovations have reached their fullest exploitations, almost any individual can shop in a grocery store a feast that many of the wealthiest kings throughout history could not have envisioned, for most part, at a reasonable price (Uchiyama, 2002). Fossil fuels are said to have insulated us from the fickleness of Mother Nature, removing the periodic famines from our lands that have cursed humanity for most of its history. With many technologies used in the preservation of food have destroyed valuable nutrient in foods, fossil fuels powered technologies have enabled humankind to not only increase the amount of their food from harvest to harvest but have also guarded the nutritional contents in the food materials. These attributes have attracted many people in relying on fossil fuels to preserve their vital food supply systems.

Fossil fuel energy has also been used in drying grain for future consumption by humankind. This energy is said to be intensive and effective because the removal of water requires large inputs of energy. For instance, in the usual drying of grains with fossil energy, the removal of one liter of water requires an average energy input of about 3,600 kcal. Fossil fuels have are being used in providing the energy needed to avail appropriate temperatures to remove water from grains for preservation purposes in the most effective and affordable strategy. In surveying corn drying in the United States, Pimentel reported an energy input of 1,520 kcal per liter of water removed from corn by use of fossil energy. Put another way, 1,525 kcal are expended to reduce the moisture level of 7.5 kg of field-harvested corn from about 26.5 percent to 13 percent all relying on fossil fuels technologies.

A tremendous amount of fossil fuel energy is also used to transport our food to various distribution sectors after production in reaching the ultimate consumer of these products. Fossil fuel energy has made it possible for the development of centralized industrial agricultural operations and the corresponding disappearance of local family farms. As a result, food is now shipped through extraordinarily long distances before it reaches our dinner plate. This has enabled many countries to specialize in what they can produce best and supply the surplus to other countries in exchange for what they do not have a comparative advantage in producing. According to the U.S. Department of Transportation, food and agricultural products including imported or exported large quantities of different varieties of foods are transported within U.S. borders each year by fossil fuels energy constituting more than 20% of total U.S. energy used in transport. In 1969, the U.S. Department of Energy estimated that, on average, food traveled about 1,346 miles (van Ruijven et al., 2019). Another study conducted later determined that fresh produce traveled almost about 1,500 miles. In addition, an increasing quantity of food is now being transported internationally, for example in 1998, a total of 173 million tons of food were shipped into and out of the U.S and in 2001. The U.S. imported about 40% of all fruits, 13% of vegetables, 42% of lamb, and 80% of fish and shellfish where about 30% of them were shipped using fossil fuel energy. However, this excessive and unnecessary food transportation requires the consumption of large quantities of fossil fuel, thus negatively polluting the environment and damaging human health.

Lengthy food transport also generates additional energy expenditures by creating the need for increased food packaging, processing, and refrigeration. It also forces the government to spend more of our tax dollars repairing, constructing, and maintaining an extensive system of roads and railways. Given the damage to human health and the environment caused by the use of fossil fuels from some studies some authors have recommended a reduction in the use of fossil fuels however possible especially to the consumers. This is done to assist in minimizing agriculture-related fossil fuel energy usage by simply choosing to buy food from available local farms. This move will reduce the distance that food is shipped, thus conserving fuel, reducing pollution, and protecting human health. Other researchers have argued that the production of food will not be effective if countries are to produce what they cannot produce at maximum using the available resources sustainably.

The food system is currently dependent on fossil fuels for powering irrigation pumps, especially in this era of climate change. This has enabled farmers to move along with their production activities using the low paid energy source in watering their plants. Others have used the fossils' source of energy in maintaining animal operations at the farm level.

Figure 7.10: Sustainable energy for pumping and irrigation.

Source: Energypedia

The fossil fuel dependence of the food system in developed countries is much larger still because farm production accounts only for about 20% of the total system inputs. Indeed, the largest single contribution comes from food preservation, processing and preparation in the home, with additional substantial contributions from transportation and food production in mechanizing farm machinery. Past studies have shown that the continued large-scale human exploitation of coal and fossil fuel energy has occurred for little more than 200 years; for oil and gas, the timescale is little more than 100 years. The evidence is accumulating that production of these non-renewable resources will reach a maximum in the coming decades during which humankind must meet the challenge of food security. Optimal production could nevertheless come much sooner. In processing, fossil fuels are used in providing the energy needed to transform raw agricultural products into more finished goods for consumption purposes. This fossil energy has enabled the developing countries to untangle themselves from the plague of exporting primary agricultural products that fetch low foreign exchanges (Wang et al., 2011).

Fossil fuel technologies applications primarily on the farm and the interaction of the farm system articulate likely impacts of rising fossil fuel costs and conceivable adaption in developed and developing countries, both on the farm and in post farm segments of the food system. Because of its annual cycle of production, the food system adapts relatively quickly to altered prices of inputs, leading to optimism that the system will evolve relatively smoothly in a changing cost environment. Price spikes, however, can be more disruptive than gradual increases, although they may also induce long-term adjustment. An example is a radical improvement in the energy efficiency of American refrigerators since the oil price shocks of the 1970s. Since then, energy efficiency has improved by a factor of about 3.6, while average sizes have grown about 11%, and real prices have decreased by 2/3 than before. The heavy fossil fuel reliance of the agricultural system in the U.S., which should be broadly typical of other developed countries, will prompt evolution towards more sustainable cultivation systems with increasing energy costs in realizing food security among all people.

Technological fossil fuels extraction in developing inputs from improved crop strains, precision application of water and nutrients into the soils for production purposes and combined with other innovations will enhance food security. Higher energy prices will result in accelerating the longstanding evolution of agriculture and the larger economy toward higher energy efficiency. Besides, they may also reverse the corresponding trend toward greater overall energy use. Because of the significant food system costs to individual households, food and energy price increases will directly affect household budgets, leading to choices of more efficient appliances and possibly dietary changes. Owing to the long life of home appliances, however, evolution will be slow. Fuel costs will directly affect the transport sector, while electricity costs will affect food processing. Possible responses are greater reliance on biofuels, which exacerbate food price increases, reliance on fossil fuel energy sources and localization of processing and distribution facilities. The evolution probably will include all three, in a variable mix depending on national and regional market forces. Optimism about the adaptability of both the food and energy systems is likely well placed unless severe economic shocks of the more pessimistic peak oil forecasts disrupt the economy on a large scale and undermine capital investment. Then, the more drastic localization scenario represented by the Cuban experience may be relevant. Agricultural and energy policy initiatives could either accelerate or inhibit adaptation to an environment of gradually increasing energy prices; they also could reduce or increase vulnerability to price shocks.

Unfortunately, politically powerful stakes in conventional agriculture and fossil fuel production are likely to oppose policy innovations to promote more sustainable systems. Additionally, developing countries have displayed the distribution of farm energy inputs in developing countries, showing that fertilizer embodies the largest energy cost. Rising prices will make this input increasingly expensive, and probably unreachable for most small farmers unless subsidized by government policy. The cost of fertilizer will exert pressure to accelerate agroecological innovations, although soil fertility already compromised by the reduction in animal nutrients and rotation cycles may warrant targeted use of synthetic fertilizer as part of a long-range plan for land restoration (Waibel et al., 2019). The scarcity of other fossil fuel inputs will confirm existing patterns of labor-intensive farming. Success in global efforts to raise crop yields on small farms in the developing world will require sustained policy commitments from national and international agencies. Little information is available about post-farm energy inputs in developing country agricultural systems. Organic systems in developed countries may provide a partial model, although these data also are few. Transportation costs will largely confine distribution to local and regional markets, although success in intensifying production may increasingly satisfy local needs and require expansion of regional marketing opportunities. Urban farming should have a role to play, as should innovative approaches that simultaneously minimize waste, generate income, and provide food, e.g., cultivating mushrooms on invasive water hyacinths in Africa. Larger population growth rates also occur in countries where agricultural fossil fuel energies and the environment productivity is low, systemically addressing the challenges in availing of fossil fuels in agricultural activities, could advance food security both by improving productivity and by enhancing farm incomes, in addition to providing broader societal benefits.

7.5.2 Renewable Powered Technologies

Renewable energy and agricultural studies have been carried out in the recent past intending to identify the major linkages existing between them. Information on the importance of using renewable energy sources in carrying out agricultural activities is now available to the farmers and other stakeholders. This clean energy is advocated as an opportunity for rural economic development through the growth in the agricultural sectors in the rural areas. Many farmers are already producing renewable energy by growing corn to make ethanol. Additionally, increasingly farmers and other institutions are adding to their incomes by harvesting the wind that blows across their land to make electricity for their household and farm activities. Other new options of recycling technology are however becoming available.

Renewable energy sources and farming activities are a successful combination. Wind and solar energy and biomass energy have been harvested since time immemorial and are still promising to provide energy for the future sustainably without depleting. As a result, these energies will provide farmers and other investors in these technologies with a long –term source of income. These renewable sources of energy have been used in the farms to replace other sources of fuels or as sold products. Past studies by the U.S. Department of energy have indicated that wind energy alone is contributing to about 80,000 employment opportunities and about $1.2 billion in additional income for farmers and rural landowners by 2020.These renewable energies will help in the efforts of reducing pollution, global warming and dependence on imported fuels, which are very expensive.

Crop production farms as they shift from traditional to modern agricultural practices are currently using wind power to pump water and generate electricity for production. Recently, wind developers have gained popularity especially in the developed countries and have installed large wind turbines on farms and ranches in a number of states to provide power to electric companies and consumers. The presence of strong winds in the area may prompt developers to pay as much as $5,000 per year for each turbine to be installed. One turbine uses less than half an acre. Farmers have the privilege of planting crops and rear livestock right at the base of the turbine. While some farmers purchase wind turbines others are starting to form wind power cooperatives to join hands in exploiting this adventurous benefiting source of energy (Weiss et al., 2010). The use of these turbines today is mainly practiced in farming and is majorly being installed in the mid-west, Great Plains and west, where the government or state policies in place provide support. Adoption of these wind technology by many farmers in their agricultural lands would provide energy from some of the best wind resources.

Biomass energy is another type of renewable energy source majorly applied in agricultural production activities. This type of energy is produced from plants and organic wastes from everything including crops, trees and crop residues to manure. There are specific crops that are grown for energy purposes mainly grown in large quantities like most food crops. An example is the corn crop, which is widely used as a source of fuel, native prairie grasses such as switch grass, or fast –growing treess such as poplar and willow are likely to become the most popular in the near future. They are perennial crops that require less maintenance and low inputs requirements compared to crops such as corn for they are cheaper and more sustainable to produce (York, 2016). The crops combined with other biomass wastes can be converted to energy on the farm or sold to energy companies that produce fuel for cars and tractors and heat and power for homes and agricultural farms.

Figure 7.11: Solar Renewable Energy in Agriculture.

Source: The Electricity Hub

Solar energy also falls under the renewable energy sources available to the agricultural sectors. It is essentially from the enormous amounts of energy from the sun that reaches the Earth every day. These energies stored in the earth including reserves from coal, oil and natural gas is equal to the energy from only 20 days of sunshine. Desert areas are at an upper hand to make solar energy practical. This energy is used in agriculture in saving money, increasing self-reliance and reducing pollution if other sources of energy were used. Solar energy reduces the farm's electricity and heating costs. The solar heat collectors are used to dry grain crops in the farm and warm homes, livestock places and greenhouses. The solar water heaters are useful in providing hot water for dairy operations in pen cleaning. Moreover, solar electric panels can power farm operations and water pumps, lights and electric fences essentially because it is less expensive than the use of other power lines.

7.5.3 Manual /Low –Tech Technologies

This is a type of technology that uses low levels of input in the farming systems intending to optimize the management and use of the available farm inputs at the farmer's disposal and minimize the use of some production inputs such as the purchased fertilizers and pesticides. Wherever and whenever feasible and practicable, to lower production costs, to avoid pollution of surface and groundwater, to reduce pesticide residues in food, to reduce a farmer's overall risk, and to increase both short- and long- term farm profitability at the farm level. Basic materials needed for the low – tech-farming systems include a sprouted plant, a growing medium, nutrient solution, and net cups (Yoon, 2018). The growing mediums entail materials like clay pebbles, perlite, and act as support for your plants in place of soil. Nutrient solutions include mixes of nutrients that are combined with water to be fed to plants, providing all the minerals they need to grow. These technologies' adoption and intensive use would lead to increased production and enhanced livelihoods for farmers.

Chapter 8

Challenges on Agriculture Energy Use

CONTENTS

8.1 INTRODUCTION

Between 1960 and 2015, agricultural production more than tripled, largely due to the productivity-enhancing technologies of the Green Revolution and the rapid extension of land, water and other natural resources for agricultural purposes. Around the same moment, there has been a phenomenal period of food and agricultural industrialization and globalization. At around the same time, the phenomenon of industrialization and the globalization of food and agriculture took place. The physical distance from farm to the platform has increased considerably, and the use of processed, packaged and frozen food goods has increased in all but the most distant parts of the world. The central issue in the future is whether agriculture and food systems will fulfill the needs of today's global population, which is projected to exceed more than nine billion by the middle of the century and to reach more than eleven billion by the end of the century.

Food and clean fresh water are needed to grow food. With demand and climate change, on the one hand, it is difficult to find the amount of freshwater required to satisfy their needs in many areas, particularly in the south. How are we going to continue to evolve without knowing genetics? Agriculture is expected to benefit from more sustainable use of water. However, water is not the only thing that needs to improve in the agriculture sector. Agriculture is responsible for a significant portion of energy consumption. Things are very critical in this region if we are to achieve our climate change goals, but some proposals are in place (Armaroli & Balzani, 2007). The global food system relies on fossil fuel for the production of basic agricultural monocultures, from the planting of fertilizers to the manufacturing of food products on the market. Food crops are increasingly used for energy use, in particular maize, as fuel for the manufacture of ethanol. The current reliance on inadequate energy sources which adversely affect the natural environment in industrial agriculture is unsustainable. Energy choices will improve the longevity of our food system at every point of the process.

80% of the energy needs of the agricultural sector are met by fossil fuels. Petroleum consumption (domestic fuel oil and hard fuel oil, diesel and petroleum) accounted for around 70 percent of the total energy consumption of the agricultural sector in 2011. Just 8% of this consumption still accounts for the consumption of natural gas. And demand has risen sharply since 1980. The proportion of green energy sources and waste management is also modest, with renewables accounting for just 4 percent of the agricultural energy sector in 2011. Most of today's agricultural items are highly processed and packaged, which further raises the oil footprint.

Power is used in the manufacture of processed food for cooking, baking and otherwise preparing the substance for grinding, cutting and mixing of the ingredients. Food packaging is often made from resources; the majority of food packaging is made from petroleum plastic. The amount of energy used in packaging varies considerably depending on the size, shape and quantity of the items. As a result of the rise and centralization of demand, food is always transported over long distances and requires more energy inputs.

In manufacturing farms, large commercial farming facilities, which raise thousands of bovine animals without access to pasture under crowded conditions, the majority of meat, eggs and dairy products are currently being processed. Factory farms employ the above-mentioned energy-intensive methods, which require an immense amount of feed from commercial crops, so livestock cannot graze. Almost half of the 90 million acres of maize planted land end up as grain. Corn has around 95 percent feed grain output. Corn makes up approximately 28 percent of all acres of farmland in the United States. Agricultural agriculture is also a potential source of soil and surface water drainage, which could force cities and wastewater plants to spend more on water management facilities and capital (Anishin et al., 2013). These energy sources are used mainly indoor heating systems for livestock. It is estimated that we will require 20 liters of cow fuel for slaughter, 3 liters of chicken and 33 liters of pork for cooking only. Both costs can be exacerbated by fuel demand and the scarcity of food for travel.

The use of farm energy is also a big part of the use of pesticides, fertilizers and herbicides. For example, they rent aircraft in the United States that are heavily polluting and dangerous for the atmosphere in order to transport their products. The remaining 18% of energy usage in the agriculture sector was electricity (mainly produced by nuclear reactors in Europe). Energy consumption in agriculture was twice as high in 2011 as in Europe in 1985. We need to improve, however, in this area. The agricultural climate, which is very important to energy spending, is one of the most creative sectors in the Netherlands. Not all inventions are, of course, technological, and much is founded on know-how and genuine natural knowledge. Any approach is naturally oriented. The creation of "smart" environments, including mosquitoes, mites and microscopic worms, seems to be very effective in the battle against dangerous insects, such as the elimination of herbicides or pesticides. This is the principle of 'biological control' that preserves plants. Biological control is one example of many others, but it appears to be a potential choice in the future if we want to eradicate pesticide contamination and pursue a solution to GMOs.

Below are the main challenges of agriculture energy use:

8.2 INCREASING ENERGY PRICE

Agricultural production, either through direct energy usage or energy-dependent inputs such as fertilizers, is vulnerable to increased energy costs. Various factors, including the difference in oil and gas rates and commodity taxes and subsidies, can influence commodity prices faced by US farmers and ranchers. As a result of carbon taxes, regulated emission limits or the implementation of a reduction in the emissions market, energy prices are also impacted by climate change policies. Here, we discuss the role of energy in agriculture and concentrate on the results of a case study on the economic effect on agriculture of energy price shifts arising from practical scenarios for reducing greenhouse gas emissions. In general, higher energy costs of production, whatever the cause of the rise in oil prices, would reduce agricultural output, raise agricultural output, and increase agricultural commodity prices and lower agricultural output. The goal of the study is to demonstrate the peculiar distribution by the price of their carbon content of the findings of market (or cost) shifts in various types of electricity.

From the study, we learn that the cost of output for different plants differs greatly with regard to energy. The highest energy prices per acre were associated with maize and rice. Soya beans used in the analysis are among the eight key plants studied (corn, sorghum, barley, oats, wheat, rice, upland cotton and soya beans). That's the bottom of it. The combined size of the eight crops will decrease by 0.2 percent (on average below the lower price) to 0.4 percent (higher price adjustments) over the 2012-18 period with higher energy prices (fuel increased by an average of 2.6 percent to 5.3 percent; fertilizer increased by 4 to 10 percent). Food suppliers also have oil-based prices. If it is real, lower oil price change conditions (0.2 to 0.6 percent per year, the average for 2012-18) and higher (0.6 to 1.3 percent higher) lower direct energy prices than grain produced and higher feed costs for farmers. Poultry, which is the most efficient feed-to-flee converter between animals, would be less affected than beef and pork in poultry production (Ahmad & Zhang, 2020). Future technical advances (other than those suggested in the comparative scenario) have not been taken into account in the scenarios analyzed in response to continued increases in energy prices.

Moreover, the decades-long downward trend is projected to continue in the agricultural sector in terms of energy consumption per unit of production, only marginalized in higher-yield scenarios. Under these circumstances, the effects of the increase in oil prices on agriculture are likely to be overestimated. On the other hand, the longer-term impact of more increases in the price of electricity would not be as strong as the resulting short-term results reported. Effects also range from country to region. The Portal area in Mississippi is most affected by higher energy costs due to the prevalence of intensive fertilizer crops such as cotton. The net cash earnings of farmers in this region would be limited.

With less and more electricity, the average price shift potential (in 2014) was 8 to 19 percent. While supermarket food prices depend on higher agricultural commodity prices, retail food prices are more influential than they are in the production of agricultural commodities as food is processed, transported and distributed. With conditions and timeframes based on this study and without simulation of energy markets for electricity, coal and natural gas, the Consumer Price Index (CPI) for both domestic and non-domestic food would be 0.6 to 0.9 percent higher. The higher oil prices scenarios studied in this study in the agriculture sector do not appear to have a significant economic and population effect. In comparison, rural districts appear to have relatively few college graduates, a disproportionately high percentage of workers and low crime rates compared to other non-Metro districts. Any agricultural-dependent county in the Portal area of Mississippi could have a comparatively greater effect on energy-related farm income losses.

Reducing fossil fuel emissions under the carbon tax or cap-and-trade scheme would minimize overall employment in the energy industries affected. Energy-generating counties are mostly agricultural. However, few non-metro countries produce a large proportion of non-farm jobs from their energy generation, with the exception of some coal counties, which are mostly located in East Kentucky and West Virginia, there would be no overall rural effects. Significant quantities of gasoline, such as direct combustion of fossil fuels or indirect energy inputs, in particular fertilizers, may be consumed by agricultural development. Agricultural productivity in 2005-08, direct agricultural investment in the USA accounted for about 6.7% of total production expenditure and 6.6% of fertilizer expenditure. The average energy content for large plants is, however, considerably improved. A variety of factors can be used to measure the energy costs of US agriculture: natural gas production, global oil markets, energy taxes or carbon dioxide reduction legislation (Basore et al., 2018).

Illustrating the transfer of agricultural energy prices to the agricultural system. We are modeling three farm-to-detail models: a land growth comparison scenario for 2012-2018 and two projections for higher oil prices over the same timeframe. The expense of greenhouse gas pollution for illustrative purposes is motivating the instability of oil prices in alternative scenarios. The analysis uses a series of models maintained by the USDA Economic Research Service (ERS). The goal of this case study is to demonstrate the use of the U.S. agricultural environment in ERS models to simulate changes in energy prices; provide two forecasts, namely carbon pricing for fossil fuels, of higher oil prices, and provide a broad overview of the process used to deliver a previous, related study. A variety of alternatives to mitigate greenhouse gas emissions have been implemented over the last 5 years and new approaches are planned. In order to build the future. Greenhouse gas pollution pricing projections and their related energy sector implications are based on, or revised by, the U.S. studies in the two main scenarios outlined in this report. Environmental Management Agency and the United States Environmental Management Authority. Regulation of knowledge of energy.

As such, the oil prices examined are part of a series of recent climate change debates and are meant to be more illustrative than anticipated. The EPA and the EIA have provided independent analyzes of the US Clean Energy and Security Act 2009 including a cap-and-trade program. However, the results listed do not reflect on limiting pollution from trade and emissions. Otherwise, the carbon tax would have driven higher energy prices. Agricultural production, especially in field crops, consumes significant amounts of energy. As a result, oil prices have an impact on agricultural production costs. Production costs are important to farmers' net returns (profitability) and are characterized as sales revenues minus production costs, and net returns have an effect on farmers' production decisions. The effect of net returns on crops by affecting the allocation of acres per season. The choices of farmers' livestock production are often made by means of net returns subject to biological limitations.

As with crude, the sector's energy consumption can be direct. For the operation or indirect use of irrigation machinery, including fertilizers, petrol, coal, gas, electricity and services. Over the period 2005-08, direct expenditure on energy consumption amounted to an average of around 6.7 percent of the overall investment in industrial production, while the cost of fertilizer amounted to another 6.6 percent. This time, oil prices grew and the share of these inputs was roughly 15% in 2008. In addition, the cost of animal feed includes indirect energy costs due to the effect on crop prices, such as maize and soya meat. As the energy output intensity, the value of oil and the impact of commodity price changes are not consistent across commodity or country. It's very distinct. The operating costs of any commodity are out-of-pocket cash for handling the inputs (Asif & Muneer, 2007). Operational costs are the quantities and values of production inputs and are often subject to the manufacturing activities of producers. Inputs such as seeds, fertilizers, fuel, lube, energy, feed, chemicals and maintenance are priced. Sorghum has the largest share of energy-related inputs, while cotton has the lowest. Power input groups account for more than 50% of the operating costs of sorghum and oats.

The energy-related cost of production for these crops is specifically assigned. In absolute terms, running costs per acre are critical in calculating net production yields that have an effect on farmers' crop choices. Barley, maize and cotton have the highest price per acre for energy-related inputs. While the cost of petrol, lube and electricity per acre is the highest for rice and cotton, maize is the highest fertilizer. Again, comparatively low energy costs are related to the processing of soya beans. Compared to lower livestock running costs, direct energy costs amount to about 4 percent of the cost of hog processing, 5 percent of the cost of milk production and approximately 10 percent of cow-calf activities, on average in 2007-08 (Batterberry et al., 2010). In addition, livestock activities have an indirect effect on energy markets due to higher feed rates. Feed costs accounted for about 11 percent of total cow-calf running costs in 2007-08, 58 percent of hog production and 76 percent of milk production.

Energy-related input costs often vary by region, largely due to crop composition and irrigation dependence. There are two markets on the opposite side of the oil market for wheat and soybeans. The most energy-intensive production costs for maize, the Fertile Rim, the Heartland and the Prairie Gateway are in the areas. Southern Seaboard and Eastern Uplands are the regions with the largest share of energy-inherent soya-beans production costs. The North Great Plains have the lowest proportion of energy sources for wheat and soybean, well below the national averages of 60% and 31%. The cost of wheat processing is an interesting contrast in running costs in the northern Great Plains and the Prairie Gateway, where most grains are grown. The prices of gas, lubrication and electricity are different in both markets, while the cost of production is similar due to the cost of fertilizers

(37 percent), (28% for the Prairie Portal, compared to 12% for the Northern Great Plains). The high level of use of irrigation in the Prairie Gateway is partly responsible.

Irrigation is a significant contributor to US agricultural production. In 2007, agricultural products sold on all farms amounted to $297.2 billion in sales value, while irrigated farms were irrigated (a farm that irrigates any land). Approximately 40% of the amount. The total agricultural value ($393,700) of irrigated agricultural goods produced was four times that of average non-irrigated agricultural products sold ($93,900). Obviously, irrigation adds to the value of plants sold, but it also contributes to the agricultural value of livestock and poultry by using irrigated plants as feed (Badawy et al., 2010). The utilization of energy in irrigated agriculture depends primarily on the amount of irrigation land, the amount of irrigation water used and the efficiency of irrigation. More irrigation schemes and treatment practices usually promise lower costs per acre. For U.S. agriculture, irrigation runoff energy costs amounted to more than $2.6 billion, again mainly in the 17 western countries in 2008. The distribution of electricity irrigation costs in this sector has a bearing on the effects of the Partial Budget on provincial agricultural enterprises.

The key costs of irrigated cultivation are the costs of irrigated pumping of electricity. However, the total water supply costs for the 17 West States for 2008 included $671.5 million in import costs, irrigation reconstruction and conservation costs of $659.6 billion and labor irrigation and contract irrigation costs of $870.0 million. Air production, mainly carbon dioxide, nitrogen and nitrous oxides, offsets around 8 percent of US greenhouse gas emissions. With the high cost of energy such as electricity, it is quite difficult for farmers to conduct agricultural activities. Therefore, some of the farmers are unable to water their crops.

Carbon dioxide is released when fossil fuels are burnt, methane is generated through enteric fermentation by livestock and manure management. There have been multiple alternatives to minimize potential US emissions of greenhouse gases. The introduction of a carbon cap-and-trade scheme is normal, with a pollution quota covering the majority of US emissions of greenhouse gases. The price which sets out the specifications for carbon allowances is the corresponding premium for greenhouse gas emissions. The Cap and Trade plans do not cover agriculture: the production of methane and nitrous oxide from animals is above the norm. However, the cap-and-trade system has an indirect impact on agriculture by the price of fossil fuels, based on the carbon content of each commodity and the price of greenhouse gases.

Proportional price changes for fossil oil would explicitly benefit from carbon pricing. Agricultural use is made of vast volumes of energy: refined oil for agricultural machinery, irrigation and other plants, natural gas indirectly by nitrogen fertilizer and coal indirectly by electricity. The price of greenhouse gas pollution has a different effect on fuel, depending on the carbon content of each energy unit. Coal has the highest carbon emissions from global fossil fuels such as carbon, refined oil and natural gas per unit of oil and natural gas. As a result, not only do carbon dioxide prices boost energy prices, but the perceived cost of fossil oil also changes. Despite the expense of producing greenhouse emissions, natural gas will be cheaper than coal and refined oil. A simple measurement unit is required to compare the effects of different greenhouse gasses on the atmosphere. Greenhouse gas compounds other than carbon dioxide are then converted to their equivalents in carbon dioxide to guarantee that total emissions are rounded up in metric tons of carbon dioxide equivalent (t CO2-eq)4 from several Greenhouse gases.

Few non-metro-regions generate a large proportion of non-farm energy-producing jobs, but improvements in rural emission rates will have a marginal effect. However, the effect on unemployment will be higher in non-metro areas, primarily because the oil mining regions are mainly rural and some counties are seriously affected. The lower levels of schooling and economic jobs in these regions make it particularly difficult for the coal mining counties of East Kentucky and West Virginia to adapt to major job losses. Further developments would lead to long-term rises in oil prices in the agricultural sector. Emerging technology introduction and rollout would foster changes in agricultural productivity and business activities and ultimately address the economic effects of higher energy prices (Cardenas et al., 2016). Farm manufacturing systems must be revised in order to improve energy efficiency. Oil consumption is projected to resume its decades-long decrease per unit of production in the agriculture sector, which would minimize impacts on energy prices as well.

8.3 DECLINE IN FERTILIZER CAPACITY PRODUCTION

As humans require vital minerals and nutrients for good, steady growth, so do plants on earth. Our major fertilizer sector is closely linked to agriculture and food production. The role of fertilizer in food production is usually underestimated. The fuel for the plants is manure. Fertilizers consume plant nutrients that are lacking in crops. Without the addition of fertilizers, grain production and output would collapse significantly. This is why soil nitrogen reserves are supplemented by minerals that plants can readily absorb and use by mineral fertilizers. We should then add to what we do to satisfy the human needs for the food and meat we consume. This combination needs to be corrected, as well as a degree of soil nitrogen conservation to protect our crops. Each field is made up of these stocks and we need to add fertilizers every year and every harvest. In other words, we use fertilizer to have surface nutrients that are not available. Missing nutrients are applied during harvesting. Align nutrients to improve productivity and production efficiency.

Agriculture is commonly referred to as economic stability, development and change, both directly and indirectly. Farm production speeds were seen as optimistic and confident for economic growth. Economic development has seen a drastic structural change in the economy, expanding on the past experience of Western countries, and relies primarily on agricultural practices. As a result, agriculture's primary purpose was to provide enough low-priced food and human strength to the expanding industrial economy as a strategic leader in all economic growth strategies. Any economic growth policy plays an invaluable role in the agricultural sector, particularly in the low-income developed countries. Agriculture has many restrictions, including soil erosion (salt salinity, alkaline, erosion and surface fertility), water scarcity, irrigation mismanagement, limited land distribution and inadequacy of farming methods. In all cases, it is expensive and inefficient to use agricultural inputs, including fertilizers. Available are limited seeds and stocks of pesticides.

Fertilizers play a vital role in helping farmers to reach their high levels of productivity. Our farmers face the biggest problem of reducing the efficiency of crop production. Weak soils affected by continuous farming without the use of adequate mineral fertilizers and dough are the main factors contributing to a decline in soil fertility. Adequate use of agricultural inputs would ensure that agricultural production remains consistently high. The other option available is seed fertilization. Fertilizers are plant-supplied chemical compounds that are typically used either from the soil or from the leaves to promote growth. There are two types of fertilizer available, called organic and inorganic fertilizers. They are both known as 'mantle.' They are essential plants nutrients provided by fertilizers, namely phosphorus (P), zinc (Zn), boron (B) and sulfur (S). Nitrogen brings an immediate green power to the farmer that the farmer believes the true impact of any productive fertilizer, while phosphorus has a significant role to play in root production, energy conversion and, finally, crop yield. Balanced fertilization is a critical approach for optimal soil growth. Fair fertilization is characterized as the fair use of fertilizers and other inputs for the best supply of all the nutrients necessary for maximum crop yield. Fertilizers are not cheap, but they need to be deliberately and correctly used to allow the best possible return on their investment. Balanced fertilization does not mean that fertilizers (nitrogen, phosphorus and potassium or any other nutrient) are applied to agricultural supplies, but must take into account the excess of nutrients already available in the soiling and crop needs and other factors. Removal, economy and efficiency of fertilizers, ability to spend for farmers, agro technology and field moisture should be taken into account.

Complete US use of nitrogen was approximately 21.4 million tons short in 2002, with GAO's claim that approximately 12 million tons (or 56%) had been used by farmers. The US industry used more than 9 million tons of nitrogen, including bacterial growth induced by waste disposal, plastic refining and cooling. Fertilizers are highly energy-intensive, particularly nitrogen fertilizers. The cost of processing nitrogen fertilizers, either directly as a feed or indirectly as the energy production fuel required for production, accounts, as has already been stated, for a considerable part of the cost of natural gas (75 to 90 percent). As domestic gas prices increase the competitiveness of US fertilizer farmers. In foreign countries with high nitrogen export capacity, prices of natural gas remain well below US prices.

For example, at the beginning of 2001, when the prices of US natural gas were high. Middle East gas prices were Btu 60 million, North Africa 40 percent, Russia 70 percent and Venezuela 50 percent, and Btu 5 million approximately. Like natural gas, the federal government does not set or track the cost of nitrogen fertilizers. In contrast, restrictions on imports by the United States, such as quotas and tariffs, do not usually extend to products purchased by other countries from nitrogen fertilizers. U.S. fertilizer manufacturers will respond to intermittent price spikes in natural gas by briefly shutting down and restarting production, while prices will fall again. In recent years, high domestic gas supplies have led to high gas prices. A significant portion of the production potential of US nitrogen will be idle and/or closed (Ceylan et al., 2004). The US ammonia plant had a production capacity of 21.4 million tons in 1998. 3.5 million tons, with an operating capacity of 16.3 million tons (76.5 percent). Decreasing demand for nitrogen means that either the utilization of nitrogen must be limited or nitrogen must be increased. Ammonia production was closed from 1998/99 to 2003/04 for an extra 1.5 million tons.

Supporters of the American Farm Bureau Federation (AFBF)-funded U.S. fertilizer industry are committed to amending U.S. laws and regulations that either encourage development in the production of natural gas or minimize the demand for energy from natural gas. They also contend that the relaxing of environmental standards for coal-fired power stations, the expansion or lifting of nuclear and hydrocarbon licenses, and the promotion of the use of pure coal technology. They advocate greater access and higher taxes on federal lands that are now off-limits for nuclear and hydrocarbons. However, a majority of environmental groups are sustainable. Energy campaigners and city pollution reduction groups are condemning these plans. In September 2003, the National Petroleum Council (NPC) published a study on Balancing Natural Gas Policy, which discussed policy alternatives for addressing the high gas price crisis. The report acknowledged the likelihood that natural gas prices would increase 'for years to come' and proposed, among other options, that the most immediate possibilities for price increases would be to save electricity and boost energy conservation.

8.4 HIGH COST OF FERTILIZERS

More natural gas costs have resulted in slightly higher prices for nitrogen fertilizers. The price of a natural gas wellhead in April 1999 was $1.90 per 1,000 cubic feet, and the short price of ammonia was $211. Two years later, in April 2001, the well-known price of natural gas soared from 138 percent to $4.52, while the price of anhydrous ammonia increased by 89 percent to $399 per ton. The cost of the whole collection of nitrogen fertilizers is very high and gives agricultural farmers little cost-saving choices, as the main component of most nitrogen fertilizers is hydro ammonia. According to the GAO, the cost of gas and fertilizer rose dramatically in 2001, when the US fertilizer industry saw a 25 percent decrease in nitrogen demand and a fall in domestic fertilizer ability. However, GAO argued that the supply of in-house nitrogen fertilizers was "was adequate to meet the demand of farmers" due to two mitigation factors. First, US nitrogen supplies grew by 43%; and second, agricultural usage of nitrogen fertilizers declined by 7%.34. Although these market shifts were helpful in keeping supply and demand under check, they did so at a much higher rate.

US fertilizer import rates have risen in recent years and almost doubled since the index first appeared in December 2007. The prices of imported fertilizer increased by 90.4% between December 2007 and October 2008. (See map 2.) The time coincided with a drastic spike in many commodity prices, which influenced the prices of commercial grain crops as well as large fertilizer products, in particular natural gas. But the US economy, like fertilizer, has undergone a general shift in business dynamics over the last few months of 2008. Between October 2008 and July 2009, the import rate of fertilizers decreased by 39.8%. Since the summer of 2009, with a few exceptions, import prices of fertilizer have slowly risen to the recent peak in the early summer of 2012. From July 2009 to June 2012, the market index for imported fertilizers rose by 93.1 percent. The import prices of fertilizers decreased by 9.5 percent between June 2012 and March 2013. Between mid-2009 and March 2013, the two key forces that historically affected the import prices of fertilizers started to shift in the opposite direction: the prices of grain crops and the prices of US natural gas (Cook, 2021).

Seed prices have increased in the last three years. The most valuable crop was the costliest, the highest field, the most fertilizer used, and the recent high price was estimated in 2012. Since 2009, domestic maize planting has grown by 12.6 percent, particularly as a result of record high prices. The use of U.S. maize fertilizers may be about 40%. Increased demand for maize for ethanol production, drought and low grain supplies are all factors that have contributed to higher prices for maize. These price pressures have arisen in combination with other long-term developments that raise the world's demand for grain, such as population growth. The development of inorganic fertilizers is a method that uses a high amount of energy to produce and transfer electricity, oil and natural gas in particular. Approximately 74% of the energy needed for fertilizers is natural gas. Increasing energy costs have already been established as the key factor of peak fertilizer prices in 2008. While the bulk of the supply of potassium and phosphorus needs to be manufactured, commercial nitrogen fertilizers are mainly provided by naturally occurring ammonia-synthesized compounds.

The ammonia production price of the Fertilizer Institute. The cost of natural gas in ammonia is between 72% and 85%. Although Nitrogen Fertilizer and Natural Gas are the largest import trading dollar volume in the import price index, decreasing US gas prices could have put downward pressure on import fertilizer prices as more nitrogen fertilizers are produced domestically. However, most domestic producers of nitrogen fertilizers have not been able to boost demand fast enough to regain the market share of fertilizer import suppliers. Increased prices of natural gas led to the closing of US nitrogen plants in the period 2002-2008, with fertilizer prices rising as natural gas prices and imports increased. Given that natural gas is an overwhelmingly foreign commodity, the price of the United States has plummeted sharply and suddenly, it can be partially explained.

New extraction technique has been used in the United States to boost the production of natural gas more efficiently than in countries manufacturing nitrogen fertilizers. Owing to the dramatic decrease of U.S. natural gas, domestic nitrogen producers have not rapidly expanded their production ability to reduce their import reliance. Despite a downturn in the prices of natural gas, the market for fertilizer imports remained high. Established stocks of US ammonia are now cost-effective and several new installations are under construction. Although the shortage of domestic supply was a factor in the increase in the import cost of fertilizers, the need for high grain prices was the key factor in prices (Cullen et al., 2011). NFI rose by 0.1 percent in the first quarter of 2013 due to an improvement of 0.2 percent, 0.1 percent in January and February, with the exception of fuels. Monthly earnings in March were marginally offset by a decline of 0.2%. An increase in food, feed and beverage costs has helped , even though the decrease in car prices partially offsets growth.

Over time, meat, food and fluids rose by 2.4%. The first quarter of 2013 was the first quarterly increase since 1.2 percent growth was announced in the first quarter of 2012. In March, fruit and vegetable prices increased by 7.7% and in February by 3.3%. The food group rose to 2.3 percent in February and 3.8 percent in March, leading to higher prices for seafood and shellfish. Tariffs on the other major classes had a reduced influence on the general flow of import prices, with the exception of gas. With the exception of oil, the price index improved by 0.2 percent for the manufacture of products and materials.

8.5 ELECTRICITY SUPPLY ISSUES

Electricity has had at least as great an impact on modern agriculture as steam or oil, because electricity is by far much more efficient than previous electricity sources. While scientific interest in the effects of electricity on plant growth, particularly after electrical lamps have been developed, the advent of an electric motor has really attracted the interest of the farming community. Due to the apparent advantages of the other, more open channels of power, rebel farmers in a variety of countries have been able to make use of this. It is believed that the first cooperatives of this kind were established in Japan in 1900, followed by similar organizations in Germany in 1901. In addition to initiating rural electrification as such, these farmers' cooperatives also provided the basis for their future development, with a significant increase in numbers. From these early days, the advancement of rural electrification, albeit necessarily slow, has steadily increased until the general public finally forced governments to look at the development of rural electrification nationally in the 1920s.

Almost all rural facilities in the more developed countries, domestic, commercial, industrial and agricultural, are now supplied with enough electricity. Although early applications of electricity have not been fully realized over the years, power and some lighting have necessarily been restricted. Bovine cakes, grain crushers, water pumps, electric motors were used to drive barn machinery and scrapers and roots. Simple power operation and low maintenance save time and work. It was not long before the electric motor began to replace the drilling, winding and other agricultural processing equipment with a mobile steam engine outside the barn. A number of electrically driven and rope-driven plowing plants have been used in the field in a number of European countries, some of which are quite large. However, these systems were not subject to time tests or competition from a mobile tractor driven by internal combustion. Until the 1920s, when economic pressures and increased dynamism of land work led to the alteration of agricultural structures as a whole, there was no significant increase in agricultural applications for electricity (Chowdhury et al., 2020). This shift is the starting point for the evolution of agriculture from labor-intensive to capital-intensive industries, based on a new intensive crop technique resulting from the development of a wide range of mechanical, electrical and electrostatic equipment. It was important to make this change.

Modern electrical applications in agriculture range from relatively easy to as complex as production applications. Preparation and rationing of animal feed included the preparation and storage of grain and grass; and a controlled environmental framework for intensive pig and poultry farming and greenhouse horticulture planting. In dairy farming, energy is equally important for feed rationing, milking and milk cooling; all of these applications are controlled automatically. Increased applications have been made for the use of computers in the agricultural management and control of automated equipment. In order to help solve weather hazards during harvest time and minimize labor requirements, the engineer and farmer combined electrically powered crop preservation and storage equipment. In many installations, grain can be harvested and dried in days rather than months at the required moisture content for long storage with electrically driven ventilators and gas or electrical heaters.

Farmers also made more reliable predictions on how they are able to run machines depending on the availability of electricity. In some areas, it is quite difficult for farmers to access resources such as electricity that would enable them to do a number of farm operations such as processing of crop produce. Owing to the poor response rate, it would be statistically difficult to calculate average farmers' damages due to power failures. However, the cost of multiple power failure losses is measured on an approximate basis. The following categories are covered by the indicative failure: costs of electrical backup facilities, loss of cattle, loss of revenue, disruption to facilities, workers' wages unable to operate, costs of restoring the farm to full service, irrigation needs to be repaired, problems with electrical fences due to pump malfunctioning. As a group, it is not possible to measure such expenditure on the grounds of negatives rather than actual financial expenditures. Situations such as lack of irrigation water or pay have also been told. A very significant proportion of farmers stayed on the farm as a means for producing self-protection. More than 3/4 of all farmers reported getting generators in the field. The importance of this for self-insurance is further discussed in case studies for poultry and milk products.

Power cuts in the Zimbabwean agricultural sector tend to undermine development and, according to local media in the Financial Gasket, have adverse effects on tobacco planting in 2015/16. Power cuts were triggered by damage and repairs during the Hwange power plant and low water levels at the Kariba power plant. Last week, ZESA Holdings released a rigorous load shedding schedule that would leave most parts of the world powerless between four and ten o'clock a day. The ZESA Survey, 414 MW, Kariba 500 MW, Harare 30 MW, Munyati 22 MW and Bulawayo 18 MW have now been generated from Hwange. When farmers became irrigation generators, the impact on agriculture of energy cuts of up to 16 hours a day will be anticipated by expanding the planted area and increasing the cost of production (De Andrade et al., 2015). Power cuts were not only caused by tobacco transplantation but also adversely impacted by the wheat crop being prepared for harvesting. When the crop was almost ready for harvesting, power cuts began to take place and farmers could not fail to observe that, due to lack of water supply, the quality and amount of the crop had been compromised.

Sensitivity to the supply of poultry results in severe harm, including from short-term failures. It requires, in essence, considerable self-protection in this area. Both farmers had their own generation of back-up strength. Farmers are more interested in energy supply issues on this market than farmers on any other market. Due to their challenges and expertise in the relationship between the energy supply industry and the manufacturers, other farmers from other sectors have studied how to mitigate energy supply problems. This susceptibility to increased energy, considering the awareness that other farmers can use as a critical input.

8.6 ISSUES WITH RENEWABLE ENERGY

Many farmers have opted to use renewable energy to cushion energy resources that are non-renewable. Some of these include using solar energy. The project projected that farmers would migrate to solar farming if they had better income prospects, thus reducing water abstraction. It did not emphasize or include practices related to the enhancement of the use of solar energy in agriculture, irrigation and on-farm applications. A rare example of the integration between solar energy, water and agriculture was the strong emphasis on the use of solar energy as a stand-alone way to reduce water use by abandoning agricultural activities. Although this proposal was created through a participatory process at the Basin Stakeholder Forum and eventually sponsored by donors from the Ministry of Water, it was not feasible to catalyze adequate support for three different purposes.

The sophistication of the variables that contributed to this transition is a key problem in minimizing the need for water from solar energy agriculture. In certain cases, this is a bold move. Perhaps the SEF in Azraq was trying to cause a long-term downstream effect. In other terms, farmers become energy contractors, land use is minimized and the key issues of the water market are taken into account as abstractions of water. This purposely discouraged SEF-activities from being connected to land and water use or farmers from using solar power. However, the long-term effects on farmers' livelihoods cannot be achieved by a strategy that implements technical solutions (pilot solar farms) or implies viability benefits (economic feasibility for farmers). This is particularly valid because there are disincentives and threats in the face of these improvements. For example, while land productivity and benefit rates (disincentives) have decreased for Azraq, there are still favorable factors in agriculture, such as high energy and water subsidies and secret incentives for illegal land use. Nothing has been shown, in contrast, of the large pull forces in the solar industry. There was no doubt that the cost of transition deals was high due to a lack of clarification of changing situations and procedures, a lack of attractive business funding mechanisms and a lack of potential viability (Delmastro & Gargiulo, 2020).

As a result, farmers were presented with the possibility of confusion about financial subsidies related to the pilot project and the usability of an attractive FIT. The threats were also correlated with the continued absorption of water and the associated costs of managing and connecting the network (e.g. offices, power plants, distribution providers, renewables) and with a low level of overall support and involvement, especially from the agricultural sector. Recent ventures in India show that their livelihoods can be partially transformed in the short term, providing that farmers expect solar farm incomes to rise steadily. In compliance with the Energy Purchase Deal, these services are often known to be a product of the energy usage program. India has a tradition of pumping systems (SPS) and solar energy systems since the early 1990s. The effect on fragile bodies of water, such as groundwater, is also troublesome. According to Bassi, the SPS strategies are beginning to shift the emphasis. They have also sought in the past to raise subsidies for SPS, low-cost electricity and fair access to groundwater. Scholars also emphasize that solar businesses, such as commercial water rights, pro-rate pricing and PLCs, are actively or indirectly adopting energy conservation programs. This current attention is due to the issues of previous programs, which have resulted in a shortage of SEF compliance protocols and overly large SEF facilities.

Since the decrease in irrigation systems and the rise in rural electricity have led farmers to use groundwater, India has mostly been irrigated by groundwater (Sharma, 2015). Agriculture is India's largest labor force, accounting for almost 42% of total employment in 2018. (The Bank of the World 2018). The Government is also promoting the use of solar energy in agriculture by providing high incentives for SPS capital. This is due to the fact that irrigation pumps account for 22 percent of overall energy demand, of which almost 20 million are diesel and agricultural, and to the rise in rural growth and income from agriculture (Agrawal and Jain, 2016). About three hundred thousand solar pumps were installed on their own in 2015-16 relative to the previous 25 years (MNRE 2016). There is a mirror to the rise in federal funding for SPS. Donors are gradually focusing on the introduction of more interconnected initiatives relating to the use of solar energy by agricultural production and water reduction.

By using solar energy, farmers also have to deal with challenges such as poor weather. This is because there are seasons where no sufficient sunlight is experienced in an area therefore they have to look for alternative methods to provide electrical energy.

8.7 SOLUTIONS TO AGRICULTURAL ENERGY PROBLEMS

With most of the fuels i.e. petroleum and other fuels being non-renewable and expensive, farmers have opted to use more natural remedies to solve the energy problems. Many farmers are already producing renewable energy by growing corn to make ethanol. The number of farmers and farms is now increasing as the wind is harvested, which blows to generate electricity all over their land. New alternatives are becoming more and more accessible. Renewable energy and agriculture are the winning combinations (Edelenbosch et al., 2020). Wind energy, solar energy and biomass can be harvested forever, providing farmers with long-term income. "According to the U.S. Ministry of Energy, wind energy alone could provide 80,000 new jobs for farmers and rural landowners and $1.2 billion in new revenue by 2020," he said. "Renewable energy can be used on farms to replace other fuels or can be sold as cash crops".

They include the following:

The Biomass Energy

Biomass energy is generated by organic waste and manure from crops, trees and crop residues. As an energy food crop, cultivated crops can be produced in large quantities. Indigenous grass plants, such as Switchgrass and fast-growing trees, such as poplars and willows, are likely to become the most popular in the future, although maize is currently the most widely used energy crop. These permanent crops require less maintenance, less input and more sustainable production than crops such as maize which are grown annually. Plants and biomass waste can be converted into agricultural energy or sold to energy companies producing cars and tractors for fuel compliant with the USA. By tripling the use of biomass for energy, the Department of Energy would generate up to USD 20 billion in new revenues for farmers and rural communities and could reduce emissions from global warming by as much as 70 million cars. New incentives for capturing these benefits are being provided by the federal government and several states.

The Electricity of the Wind

Wind power farms have been pumping water and generating electricity for a long time. Large wind turbines have recently been installed for power supply companies and consumers on farms and ranches in several countries by means of wind turbines. Where strong winds exist, developing companies are entitled to pay up to $2,000 to $5,000 annually for each installed turbine. Farmers can plant crops and grazed livestock right up to the base of the turbine with less than half an acre of each turbine used. Some farmers have also purchased wind turbines, while others are starting to develop wind energy cooperatives. With the support of government policies, most of the large turbines in the Midwest, Great Plains and West are now being installed. In many more countries, however, farmers could benefit from the fact that agricultural land has some of the best wind resources.

Supplying solar electricity

Every day, there is a tremendous amount of energy coming from the sun to the earth. All the energy stored in Earth's coal, oil, and natural gas reserves is only 20 days of sunlight. Although desert areas such as Arizona and Nevada are sunnier in the United States than elsewhere, most regions are sunny enough to provide solar power. Solar energy can be used in a number of ways to save money, improve self-sufficiency and reduce agricultural pollution. Solar electricity can reduce farm electricity and heating costs. Solar heat collectors can be used for dry plants and warm homes, livestock and greenhouses. Solar water heaters are capable of providing hot water for milk, plumbing and home use. Photovoltaics can be used for farms, remote water pumps, lights and fences (solar electric panels). Buildings and barns can be renovated to capture natural daylight instead of using electrical lights. Solar electricity is often less expensive than the extension of power lines.

Chapter 9

Policies and Regulations for the Energy-Agriculture

CONTENTS

9.1 INTRODUCTION

The demand for food and energy across the globe is increasing very quickly because of the rise in population as well as the rise of income in different people. However, it is important to note that food production and its expansion are limited because of various reasons that have proved to be a problem in the field of agriculture and especially energy agriculture. Some of the factors that have led to the limiting of expansion of food production include degradation of land which has rendered many places unfit for any agricultural practices. There is the issue of climate change which has proved to be a problem, especially to farmers living in different areas across the globe. There is a decreasing growth rate in productivity when it comes to agricultural activities that have led to the limiting of expansion of food production (Elghawi & El-Osta, 2015).

In order to reduce the problems caused by global warming and adverse climatic conditions, it is important to reduce the carbon emissions which are used in fossil fuels and the production of agricultural foodstuffs. In order to eliminate carbon emission, it is important to switch to renewable sources of energy which are considered to be cleaner and use resources that are conserving and those that have proved to be efficient in agricultural practices. In doing so it will allow the climate condition to favor agricultural practices as well as ensure that the agricultural sector in different countries across the world is able to expand and increase food production in different areas.

It is important to note that the deployment of massive sources of renewable energy is highly required in rural areas where there are agricultural value chains. Most of the farming activities and practices are usually done in rural areas, therefore, it is important to introduce the farmers to renewable sources of energy that will help preserve the environment especially the climate and soil which are highly considered and important when it comes to practicing agriculture. Energy is usually used directly for the introduction of agricultural goods by different farmers across the globe. The energy is used for irrigation purposes in areas where rainfall is considered to be minimal and cannot be able to sustain the normal agricultural practices which are required for there to be enough food production. It is also used in crop cultivation mostly in greenhouses where the temperatures are supposed to be regulated on a daily basis and where the plants are supposed to be monitored with utmost keenness. The energy is also used to pump groundwater which will facilitate the growth of agricultural produce. The renewable source of energy can also be used in mechanized agriculture and during the harvesting process where water is mostly required to ensure that the harvesting process has been completed and is successful. This source of energy can also be used in the transportation of agricultural produce from the farm where the farmers grow these farm produce to various markets so that they can be able to reach various consumers.

In fact, the energy footprint of global food production value chains is substantial and account for about 30% of the total global energy that is used during the farming process with a significant positive impact on the ecosystem. The crops and the residues that are left after the agricultural process has been completed are also used to produce the energy that can be used to facilitate other agricultural practices. The activities that are involved during the energy and food production activities are often competing for land which is scarce. They also compete for different materials such as water which is considered to be very scarce because of the climatic conditions that are experienced in different agricultural regions across the world. The other factories labor is hard to come across because most of the people who practice farming do not have the required skills and therefore end up creating a mess that affects food production. Most of the farmers lack enough capital resources that they need in order to facilitate their day-to-day activities that involve farming (Ersel et al., 2004). Fortunately, the Nexus between energy and agriculture is not only that of trade-offs they are also ample opportunities for synergies. Luckily, the network among energy and horticulture isn't just that of compromises, there are likewise abundant freedoms for cooperative energies. For instance, if smallholder cultivating families approach cleaner energies for cooking, they could utilize animal fertilizer for treating fields, as opposed to as fuel, and acquire higher harvest yields. Another model is if farmers approach clean cooling advances for their produce (for example milk, organic products), utilizing biogas or sunlight-based boards, post-collect damage could be diminished, improving item quality and farmer livelihoods. Utilizing such open doors in the Energy-Agriculture Nexus can, in this manner, take into account raising horticultural profitability, earnings, and thus, improve food security. Besides, the utilization of inexhaustible and clean fuel sources will help in decarbonizing the worldwide energy blend, remembering for farming, in this way, adding to environmental change relief.

To wrap things up, better admittance to energy can likewise invigorate the extension of beneficial employments of energy for country development, through growing rural and non-farm pay producing exercises by assisting with making new little and medium-sized organizations along the rural worth chains and furthermore through re-area of assembling enterprises into rural regions with more great admittance to land and work assets, consequently empowering an expansive provincial turn of events.

9.2 POLICIES AND REGULATION FOR RENEWABLE ENERGY

As stated the renewable sources of energy have a great potential for improving the income and sustainability when it comes to the agricultural value chains. However, it is important to note that such potential is usually not utilized because of lack of sufficient political powers which will challenge the technology that is based on fossil fuel. The political economy has a very vital role when it comes to developing the sector for renewable sources of energy. Policies and regulations are important when it comes to ensuring that the agricultural practices are protected and people can continue with their practices as they pay close attention to the environment and ensure that they protect the environment at all costs. During the early stages, renewable energy technologies usually lack commercial skills that are large and enabling the policies and regulation is important when it comes to promoting such renewable sources of energy. Take Brazil as an example its success in bioenergy is a result of the policies and regulations which promote the production of biofuel which is used by many farmers in agricultural practices. It is important to note that there are various sensitive issues that are considered to be political in the policies and regulation for renewable sources of energy which can include a number of things such as ensuring that there is food security even when the climatic conditions do not seem to favor agricultural practices (Esen, 2016).

The other sensitive issue is the creation of job opportunities which has proven to be difficult especially when it comes to providing jobs for people who are in the farming or other agricultural sector. The other issue is reducing the dependence of many farmers on the use of fossil fuel that are responsible for damaging the environment because of the kind of pollution that is associated with such fuel. The mitigation of climate change is also an issue that is involved in energy policies and regulation because it is hard to predict the weather in an environment that is already polluted by fossil fuel. Another issue preserving the ecological integrity and concerns over a large-scale acquisition in different countries and many more.

Bearing in mind all these factors there are two ways in which regulation and policies and energy agriculture can be viewed. The first approach that is considered to be the legalistic approach is usually considered to consist of laws rules and decreased by the government and all levels of the government as well as organizations that are considered to be non-governmental which have been vested with the power to create and amend policies and other regulations. Such regulations usually have objectives and the major objective or most of these policies is targeting achieving efficiency in energy provision, ensuring that the energy prices are fair, ensuring that everyone can be able to access the type of renewable sources of energy that are required as well as ensuring that there is environmental sustainability. The role of such regulation is to create conditions that will ensure markets continue to function efficiently. Risks associated with government deficiencies are also present when attempting to address complicated issues of resource distribution of renewable energy, which calls for market usage and consistent incentives and standards. At the same time, to resolve market problems, government policy is required. Therefore, the introduction of ambitious clean energy programs calls for constructive governance.

These instruments are categorized into different divisions under various classifications. We will split them into regulation-based ones here for convenience: Mandates and goals on clean energies, Supply-in-tariffs, Net metering and metering of access to the flexible grid; Incentive-based: and reductions in taxation, subsidies, subventions and transitions, and gentle credits. Among the above-mentioned policy tools, payments and discounts, fiscal benefits, grants and soft loans are currently being used more broadly in a variety of countries to support renewable energy in the agriculture sector. For example, through numerous national plans and programs, such as the National Rural Biogas Construction Plan (2006-2010), the Development Plan for the Agricultural Bioenergy Industry (2007-2015), which includes various subsidies and fiscal incentives, China is a prime example of a country that is strongly encouraging biogas production. The United States offers producer grants to farmers wishing to establish the production of solar energy on their farms. The Rural Electrification Fund was founded by Ethiopia to encourage the implementation of off-grid renewable energy in rural areas, and since 2010, import duties on renewable energy equipment have also been exempted. Foreign donor grants are significant outlets for the promotion of green energy in rural areas and agricultural supply chains in many developed countries.

Increasing the quality of energy usage and encouraging the transition to renewable energy sources are the key goals of energy use control. There are widely complex and various initiatives that support energy conservation.

A large portion of these steps is legislation aimed at increasing the energy efficiency of the residential sector. The policy instruments used include eco-labels that certify the environmental friendliness of various consumer and industrial goods, the creation of more energy-friendly building codes, various incentives for higher energy efficiency retrofitting of existing buildings, including heat insulation in farm buildings, more efficient lighting, ventilation, cooling and cooling systems, national initiatives to encourage constructive improvements in attitude and minimize energy waste (Ghalehkhondabi et al., 2017).

There are also vast prospects for enhancing agriculture's energy quality, provided that agricultural supply chains are big energy users. In addition to more energy-efficient farm buildings, for example, there is considerable potential for reducing energy use in crop production by efficiency steps such as zero tillage, which decreases the quantity of fuel used, and precision farming, which allows applying the exact quantities of fertilizer needed for each cropland. Regulations restricting polluting energy sources such as coal, setting quotas on carbon emissions, developing caps and trading structures, implementing environmental taxes are macroeconomic and sectoral policies supporting the transition to renewable energies. The switch to renewable energy, on the other hand, is not just a phase at the global or national level. The stakes of using renewable resources are not less, but they are potentially much more important at the level of household and society. Two methods can be pursued by the domestic transition to safer and more efficient energy sources: energy ladder or energy piling.

The member countries of UEMOA have been able to provide essential chances for agricultural- and forest-based bioenergy products and by-products. There are, in total, three main demand drivers for this region's bioenergy:

- Rising energy stability by diversification
- The Energy Mix of a Nation, reducing dependence upon imports of fossil fuels, hedging toward the price of fossil fuels vulnerability.
- Counter climate change
- Economic and agricultural stimulation growth, including the formation of markets, sectors of manufacturing, generation employment and wages, and poverty elimination.

UEMOA member-country states, working with private-private governments, the sector and members of civil society, building on the knowledge of civil society, have an important role to play in other nations. In order to understand the capacity for bioenergy in the UEMOA, to ensure political commitment, it would be important to ensure efficient policy and legislative growth and construction structures. These mechanisms are urgently needed not only to direct government act via stronger interpectoral coherence, but also to give local and international signals to the right signals producers, lenders, supporters, and the global economy (Grunewald & Diakonova, 2018). There is a range of policies and initiatives for renewable fuels, including bioenergy, that are En route worldwide. Private investment in the bioenergy sector and financial support for public and/or private investors from global, multilateral and bilateral sources may be promoted by policy and regulatory initiatives. These perspectives can be used by UEMOA policymakers as history and 'lessons learnt' in the formulation of bioenergy policies and programs for member countries. The establishment of national policy structures can be supported by regional agreements to encourage the creation and use of bioenergy.

Almost all bioenergy strategies and regulations worldwide have so far centered on first-generation biofuels that are commercially viable. Overall, policies placed in place internationally have had a huge effect on the pace and extent of despite design and installation concerns, biomass/renewable energy production. To date, practice underscores that when expanding bioenergy use, policies play a key role in success or failure. The right strategies supply or facilitate the following:

- Producers, investors, and customers are expected to have faith that competition and a return on investment will occur. They should always have a positive mindset which will enable them to achieve more and ensure that they are prepared for what is to come. This can be categorized as predictability, regularity, and durability. These policies also ensure that the energy used is durable and does not cause a lot of harm to the environment.
- Consumers would see the new goods as public recognition and encouragement, local producers must share in the benefits by policies that protect access to land and ensure inputs are equal to or better than the alternative, and open and affordable.
- Manufacturing and efficiency and improving expertise in market strategy. Pilot projects that illustrate how bioenergy and bioenergy can best be generated and processed explain the economic prospects for small farmers and processors. These Programs must develop human capital to improve agriculture at the neighborhood level.

- Where the policies are implemented there should be officials who are committed to ensuring that they serve their market. The officials should be from all levels and they should be able to address the concerns of people who are interested in this type of energy. They should be professional to ensure that they deliver what is expected of them both professionally and morally.
- The demand for products should be sustained as well as it should be sustainable. What plays a major role in the development of bioenergy is the arrangements that are established to ensure that risks are avoided, the incentives by producers and smart subsidies. However, it is important to note that the production of agricultural produce on long-term basis is usually dependent on the demand and supply which in this case should be predictable.
- The policies in different parts of the world should be coherent. This means that in order for bioenergy to become a big deal in the modern energy world policies that are both national and regional schedule support each other and ensure that whatever they stand for is whatever the other countries, especially in West Africa stand for. The policies that should be common include the feed-in-tariffs we should be equal all over the world, the rate of diesel and gasoline blending and new standards will be vital in creating markets that are predictable which will in exchange and expand employment.
- There should be an investment both private and public. For private investors, they should produce and ensure that people have equal capital in order to facilitate such practices and activities. For the public sector, they should be responsible for providing information as well as training people on bioenergy and supporting research that ensures that this type of energy is used in agricultural practices. There is a certain agency in Brazil that is specifically made for agricultural research known as EMBRAPA and it is accredited with developing plant varieties that are new as well as working with farmers in different regions across the world with an aim of identifying crops that are considered to be the best and those that will be beneficial to farmers. There is an original rice research program that is located in West Africa and is considered to be successful as it offers a model which encourages research and the development of bioenergy. It is also important to note that technologies that are meant for conversion should be used together in order to maximize their efficiency.

- Lastly, those in charge should be transparent and should ensure that the regulations are clear as well as ensure that the procedures they implement do not use a lot of capital because not all farmers will be able to access I know of the polices as well as regulations that are set by different governing bodies. It is also important to note that different governments usually support bioenergy that is specifically made for the supply and the demand side.

9.3 PROMOTE EASE OF BUSINESS OPERATION

In order for a business to become successful then the person running it should consider predicting the accurate costs of different materials that they will need in order to facilitate their business. This is important especially when it comes to people who have small businesses because they cannot afford to incur any accepted cost during their line of work. For example, it was noted that imported products that are used in agriculture are usually very expensive and they take a long time before reaching a person who had ordered it. This challenge has proven to be very costly and time-consuming when it comes to getting the products to the market and this has led to the Farmers being affected. Another negative aspect of search is that it prevents other companies from entering or operating in markets that prevents the entire population from benefiting from such services. The following are the policy areas that companies feel like they are limited which affects their ability to conduct their businesses (Goiri et al., 2015).

Clarifying of customs and tariffs: A common challenge that most innovators usually experience in the line of work is that there are inconsistent customs and tariffs especially when they are exporting and importing goods from other countries. There is also the issue of a lot of paperwork which usually affects or rather it delays their products at the port which increases the storage fee that the innovators and especially the farmers need to incur. There is also the issue of the value-added tax which usually affects most farmers as they are expected to pay a lot of money for them to be able to access the products that I ordered from other countries. This affects many farmers across the world, especially in Africa like Tanzania where products are usually imposed high taxes which becomes a problem as it hinders the development of Agriculture in such places. It has been noted that for our products to be imported with is then a person is supposed to classify it as an energy agriculture appliance which will result in easing importation operations.

These models represent that organizations regularly need current information on the best way to explore customs applications and depend on vehicle specialists to liaise with specialists and exhort on duties. Moreover, while tax and obligation systems for clean energy-farming items may exist, they are not generally comprehended or precisely applied.

Facilitate foreign Exchange and Fund access: The way to obtain currency to purchase a product and buy product inventory is a key problem for many businesses. Such foreign services and instead liquidate them in order to repay foreign buyers. The central bank company activities can be significantly impacted by legislation and lengthy procedures. For instance, serving in Papua New Guinea, the agricultural operating system noted that it took many months to transfer capital into the nation for operations both local currency including back into foreign currency for goods obtaining procurement. This, in relation to the volatility of currencies as well as tariffs of solar Photovoltaic products, harmed activities that contributed to both the organization going for Kenya. One more applicant, a cool processing facility the Nigerian business noted that the conversion and translation of currency towards Nigerian naira became complex and protracted. Delays such as these will have a potential effect on market development and distribution of commodities to customers.

It is, therefore, necessary to recognize policies to regulate exposure to international investments. The laws of this type are in where to defend and guarantee that local stores operate, but also have a detrimental effect on the burgeoning Märkten. The economies any visionary working within Tanzania, for instance, observed that an expenditure that surpassed half of the valuation of the company prompted a federal monopoly check. The agency's significant resources and five to eight periods of preparation were required to collaborate with attorneys on reporting mandatory documents.

Remove The Credit Policies Including Administrative Hurdles: Government agencies restrict loans to foster investment periods and defend borrowers against unethical actions and professions. Federal regulations, though, may also block the movement or raise capital costs. To businesses that aim to provide disadvantaged minority consumers with creative funding solutions. Certain prime features include:

- *Limits on borrowing costs.* By eliminating the scope of accounting, interest amount limits will curtail recognized as an expense. For the Kenyan Government enacted a 14.5 % limit and a 7.35 % interest margin in 2016, for starters. Level on the borrowing costs paid by banking firms recorded in Kenya on industrial lending. The goal was to decrease interest rates and eventually increase access to loans. Nonetheless, the limit unknowingly triggered a deterioration of repayment performance for microeconomic, mid-, and medium-sized companies, in the fields of commerce and farming in general. Financial institutions have gradually declined extremely risky personal loans, solicitations. For instance, five out of 10 refusals were cited by one big corporation.

- *Regulating financial services.* A few visionaries have cited the Microcredit Law of Tanzania, which governs small-scale lending pushes firms by becoming microfinance institutions to fund sustainable funding. Energy-farming goods to consumers (or off services). The Bill releases new models micro financiers' testing, enrollment, and investment criteria. A micro financier, for instance, the Banks of Mozambique is unable to start or stop a workplace without any of the express authorization of both the Banks of Tanzania. The Banks of Mozambique also has the authority to terminate microcredit permits if a company abuses the permit or the substance of the license. A renewable drip irrigation office is located in Mozambique that integrates finished financing.
- The duties charged or the mobile banking laws. A visionary in Zimbabwe renewable storage tanks observed the latest per each point taxes charged on cell phones cash settlements have prevented clients from paying the use of digital payments in remuneration payments (PAYGO) via purchases. The Capital of the U.S the Economic Feasibility has reported between 10 and 15 decreases of wireless cash transaction percentage per consumer amounts throughout the SHS sector as either tax's outcome. Likewise, several nations in West Africa sees PAYGO businesses as being electrical services, who say they are offering a power kilowatt instead of a consumer product, which should thus be immune to legislation. As a result, businesses are prohibited from operating or using certain funding models in some regions.

9.4 ENCOURAGE GROWTH IN THE SECTOR

Some implementing process to model demand growth throughout the farming and power industries by the implementation of interventions to promote access to resources and services for customers. The U.S. Department of Agriculture (USDA) issues close to $30 billion of debt, for starters. In account for both crop production and price volatility, and crop insurance for farmers. Even, USDA specifically, offers business assistance loans to farmers earning less than a fixed annual sum, enabling beneficiaries to use their goods as collateral. Likewise, the Act of Rural Electrification of 1936 allowed the USDA to include rural cooperatives, urban cooperatives, direct loans and guarantees. As a measure, the percentage of electric-powered farms over 20 years has skyrocketed between 10 days to 90 days. This was primarily attributed to 1936 Rural Electrification Act. There are also several interventions in the renewable energy-agriculture intersection that can boost the market. Centered on the challenges and sophistication of business players. Innovators give priority to interventions that tackling the four major challenges: high inventory slightly higher compared prices, higher-priced costs, insufficient knowledge of technologies, and restricted access to finance. Lower Responsibilities, as well as offer rate reductions acknowledging customs & tariffs, is crucial to ensuring the entry of goods into economies, especially when the application has not been determined yet. The cost of many renewable energy-agricultural goods is many thousands of dollars, taking them out of the control of marginal farmers in rural areas (Hamlet et al., 2010). Tariffs and VAT are capable of reflecting an important proportion of the cost of the product to end consumers. For instance, off-grid fees and import tariffs on solar goods can account for up to 40 percent of the overall cost of products. Report from the Schatz Energy Research Center and social enterprise 60 Decibels reports that customs tariffs on imports are in fact, over 30 percent of the overall device expense of off-grid refrigerators is paid for by VAT. Sub-Saharan Africa, Asia, and South America for customers.

9.5 TO REDUCE END USER-COSTS BUILD SMART SUBSIDY PROGRAMS

Quality product adoption as the client base grows from early investors to mass clients the main barrier is cost. Price control for renewable energy: agricultural technologies, factories and farmers while providing their services efficiently, dealers must expand the amount of goods they market. Nonetheless, many firms are currently evolving their corporate strategies and product ranges. Additionally, tasks in several market segments, especially during the last neighborhoods due to restricted supply chains, are costly. Spare pieces, and, among other considerations, maintenance facilities. A few innovators noticed that they were unable to do so. A few pioneers noticed that they were unable to do so. Achieving a value fact that makes everyone's goods fully available to the demand of small - scale farmers strive to serve. Subsidization and perhaps other relatively low strategies as firms work to improve organizational in setting up such markets, productivity by economies of scale may play a key role. In certain examples, governments can also opt to use long-term subsidies on goods that offer multifaceted solutions. Impacts on growth if price permanently excludes these goods from the reach of the target market.

In 2019, for instance, the government of India announced the Kian Urja Suraksha evam Utthaan evam the Mahabhiyan (KUSUM) scheme, which devotes roughly $4.7 billion to low carbon farming, initiatives. A critical part of this entails the subsidization of 17,500 off-grid solar water pumps and promoting the transfer of 10,000 grid-bound emergency generators to solar Energy to lower the amount of small-scale farmer's reliance on petrol engine motors as well as adding to the renewable energy priorities of India. KUSUM talks about sustainable development problems of conventional ready incentives are resolved through encouraging delivery firms to as well as encourage the capacity in scale shipment of goods. Started campaigning to remove taxes for solar Energy products. Exclusions from tariffs and duties which aim often, the agricultural industry is difficult to execute. It's indeed pretty straightforward. Give a tax deduction or a limited exclusion of the solar array restriction. Nonetheless, things battery packs, generators, including engines, such as it is not expected for use exclusively. Presumably, for agriculture purposes simplifying the exemption from taxes requesting. Concerning tailored waivers, verification will need to take place, (such as refrigerators that need to be used for the aims of agriculture).

9.6 ENHANCE KNOWLEDGE OF THE SECTOR

Pioneers remember why policymakers will offer somewhat assistance in building consumer visibility. Renewable energy methods offer their advantages. The financial enterprise is in emerging economies, farmers are also assigned to learn about the benefits of industrialization, renewable energy technology, communications devices, the financial business, et cetera. Governments can provide that consumer knowledge quite plausibly and efficiently. For producers, the general welfare. In addition, innovators invited knowledge at the strategy panel discussion function. Advertisements to shield customers against fake items support the change from petrol to washing more broadly, solutions help customers make better choices.

9.7 EXPAND COVERAGE OF BUSINESSES INCLUDING APPLICATION DEVELOPERS OF FUND

As retail sales grow and the industry matures, firms are pursuing long-term, financial capital. To maintain growth and to present additional difficulties. Several problems emerge or escalate as industries grow, considerably. Damage with Forex. Companies also lease or procure foreign cash assets but are charged in domestic currency. It will represent a significant forex danger in certain nations. Some visionaries have noted that due to their creativity it's unlikely for both of them to protect from currency threats on a small scale, resulting in volatility that will considerably raise sales costs. As a result, exchange rate borrowing could mitigate the obstacle (Hossain et al., 2017). Many of these challenges include legislation and regulatory structures and their respective frameworks. In the countries where they work, deployment. Challenges in policies and the rules include: duties and tariffs, exposure to and growth in foreign resources, access to credit, the right to lend and use mobile money to consumers, and the ability to distinguish goods of high quality from those of poor quality.

Chapter

10 Conservation of Agricultural Energy

CONTENTS

10.1 INTRODUCTION

Agriculture has and will always be the soul of almost all life forms in the world. From human beings to animals, be it domestic or wild animals, they all rely on agriculture for their survival.

During the first ages of evolution, the human race was not that evolved to the extent of thinking of practicing agriculture, just like animals, they would feed on several wild plants with the hope that they will not harm them in any way. As time evolved from one age to the other, they managed to distinguish the edible plants from the non-edible plants; and with time, they thought of ways of producing the edible plants in plenty, and that is how to crop-growing came into place (Armaroli&Balzani, 2007).

Since agriculture involves both the rearing of domestic animals and the growing of crops; the human race, according to evolution theories, were originally apes, and apes are known to live in forests. This fact better explains the reasons why, to date, the human race has managed to tame some animals at their homes. There exist several stories on how domestic animals, who are believed to have been wild animals, came to live with the human race in harmony.

With the change in evolution ages, change was fully incorporated into human life, with them practicing both the planting of crops and rearing of animals. Most of the domestic animals produce products that are of importance to human beings, either by serving the purpose of being good to them or bringing some sort of income by being sold to manufacture other products. As for the crops, not all of them are grown for the purposes of food, some, just like the produce obtained from domestic animals, are grown as cash crops whereby they are sold for the purposes of manufacturing products that are either edibles or non-edible.

In Agriculture, there exist several types, depending on the specialty an individual chooses or on the practices carried out in their farmlands. For instance, some individuals purely major in the rearing of animals whereas others major in purely the growth of either food or cash crops and with other individuals, practicing both the growing of crops and the rearing of animals.

10.2 AGRICULTURAL ENERGY

Due to the millions of activities that have to be carried out in farms, depending on the type of agriculture an individual decides to specialize in, a lot of energy is applied for the proper execution of the different activities.

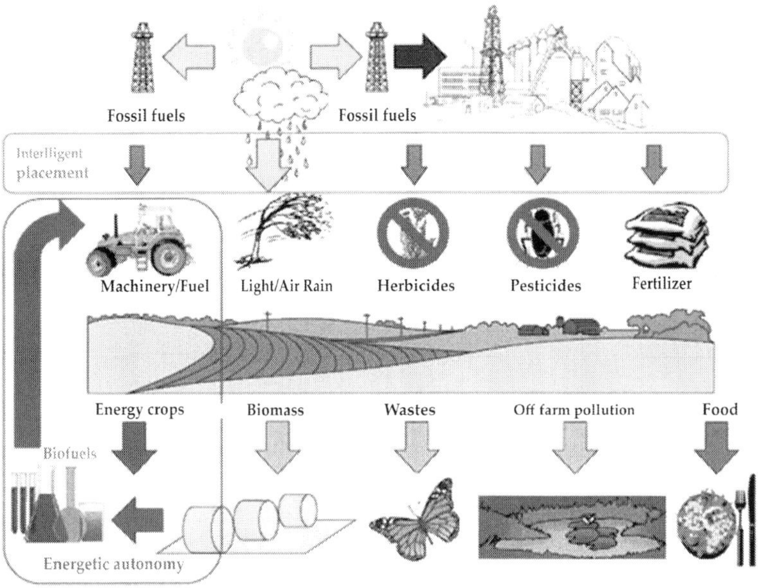

Figure 10.1: An image showing the flow of Agricultural Energy.

Source: https://www.google.com/search?q=Agricultural+energy&client=ms-android-transsion-infinix-rev1&prmd=inv&source=lnms&tbm=isch&sa=X&ved=2ahUKEwiHzsO57unuAhWLQjABHWOyBaoQ_AUoAXoECAIQAQ&biw=360&bih=592&dpr=2#imgrc=ZnpENeShPun5BM

As understood by many, energy is the amount of effort or workforce applied per unit area; that is, the total amount of labor required in a piece of land of a certain unit area, to carry out all the necessary activities aimed at improving the productivity of the piece of land.

The energy applied in farmlands can either be through human labor or mechanization of farm activities which requires some force to push the machines to do the work. When it comes to the mechanization of farm activities, the issue to do with energy comes into play at a bigger angle.

Different types of energy are applied in farmlands to ensure that activities are carried out efficiently. However, the most preferable sources of energy used in farmlands are renewable sources of energy for they save the farmers a large percentage of the cost required to ensure that they are available.

Some of the renewable sources of energy used are the likes of solar energy, which is becoming common in most farmlands; wind energy; geothermal energy and biomass energy.

As for the renewable sources of energy, they are most utilized depending on the region where the farmland is located. For instance, biomass sources of energy are mostly used in rural areas for they are easily available in large quantities, more so for the farms that specializes in the rearing of domestic animals. With biomass, the type of energy produced is known as thermal energy; meaning that, as the name suggests, the energy is produced under high-temperature conditions (Anishin et al., 2013).

Solar energy is also common in rural areas because of the high temperatures experienced in most rural areas. Hydropower is not that common in rural areas for it is a bit complex to set up; however, with the help of governments or non-governmental organizations, hydropower stations can be set in the small streams in rural areas so as to supply energy to farmers within that locality.

Some of the farm activities which benefit from the energy availed in most farmlands include; the farm water system for it requires energy so as to distribute water to different sections of the farmland; milking machines for they ought to be powered for them to function properly; refrigerators which require energy to power them; lighting systems in farmland, more so the large scale farms where farm activities are carried out during both day and night hours; among other activities.

As stated earlier, renewable sources are mostly preferred in most farmlands because they are easily available hence minimal cost. Other reasons for their preference for other sources of energy is that; they provide clean and greener energy that does not end up polluting the environment; farm business risk reduced and balancing of farm biodiversity.

The main benefit of having energy sources in farmland is to help in making the farm activities, which have been stated earlier, easier for the farmer. With renewable energy, farmers are guaranteed that their irrigation processes, greenhouse conditioning, food and feed processing and preservation are carried out within the farmland and with minimal costs being incurred.

Renewable energy sources do also boost the micro-economies of farmers in rural areas, for activities that required a lot of labor are completed within the shortest time possible.

With renewable energy sources, fossil fuels that end up polluting the environment are replaced with energy sources that boost and preserve our biodiversity.

As the Agricultural sector continues to advance, the main focus is on the provision and implementation of energy sources that are of benefit to both the farmer and the environment.

10.3 INTRODUCTION TO CONSERVATION OF AGRICULTURAL ENERGY

To ensure that the energy being used in the agricultural sector is effectively used, conservation measures have to be put into place; both for the farmer, for it, reduces the cost incurred in availing the energy within the farmland, and also to the environment, for the environment is preserved if energy is used efficiently.

To better understand the conservation of agricultural energy, it is best we first distinguish the terms efficiency and conservation. With efficiency, the work done per yield of a unit of energy is increased, that is, minimal energy is used in a large unit area, with a lot of work being done on that area. As for conservation, it basically means the reduction of the total amount of energy applied per unit area.

As stated earlier, renewable sources of energy are the most preferred sources of energy in most regions; however, that does not mean other sources of energy are not used. Be it renewable energy or non-renewable energy, all sources of energy ought to be conserved and used efficiently to reduce the cost incurred in Agricultural activities (Ahmad & Zhang, 2020).

Actually, the volatility rate of non-renewable energy sources has been the main reason why most farmers have opted for the innovation of ways that ensure reduction of the energy consumption rate per unit areas in farmland.

Some of the ways used to conserve energy in most farmlands include: proper maintenance of farm equipment which ensures that they function properly using minimal amounts of energy; the improvement of the buildings in most farmlands which ensures that minimal energy is required for farm building repair purposes; installation of high-efficiency lighting system in farmlands which ensures that most farmlands use minimal amounts of energy to power the lighting system; among other practices.

When it comes to the heating system and refrigeration system in farmlands, farmers are advised to use the latest technologies which use minimal energy to heat water for farm uses or to refrigerate farm produce ensuring that it is well preserved. Since the storage of most farm products require temperatures of certain levels, farmers could incorporate the use of sealing materials around the exterior of the storage location that reduce air infiltration hence warm conditions. All exterior walls, doors and windows can also be insulated to reduce air infiltration.

Most farmlands that specialize in the use of greenhouse operations, tend to use large amounts of energy in carrying out several activities. However, farmers could practice the use of simple techniques which reduce the total amount of energy required in greenhouses. For instance, alternative fuels such as the waste vegetable oil used in farmhouses, shell corn oil from maize plants, use woods could be used hence eliminating the consumption of fossil fuels that tends to be quite high increasing energy levels.

As for the housing of livestock, and the carrying out of different livestock operations, farmers could opt for the construction of simple livestock houses which limit the energy needed to house them and also raise them.

Field equipment such as the likes of tractors should be used efficiently so as to ensure that they require a minimal amount of energy for them to function properly. The misuse of such field equipment increases their fuel consumption rates which in turn increases the energy consumed per unit area hence increased the cost of the activities they conduct.

When it comes to irrigation, the equipment uses for irrigation ought to be changed frequently to ensure that they do not rust increasing the amount of energy used. Pumps and motor equipment should be well maintained through frequent check-ups of their components to ensure that they are in good condition. An irrigation management system should be put in place to ensure that the progress of the irrigated crops is observed so as to avoid unnecessary irritation which leads to wastage of energy that could have been used somewhere else.

The grain drying system should be improved through the innovation of a better system that uses minimal amounts of energy for the drying of grains and takes the shortest time possible to drain the grains hence improved energy efficiency.

Activities such as improved grazing activities can improve the conservation of energy. For instance, through zero-grazing, the waste from livestock is easily gathered and can later be recycled and used as manure in farmlands reducing the cost incurred in purchasing artificial fertilizers and the energy required in the application of the fertilizers for the ought to be applied in a specific amount, at specific periods of the growth phase of a plant.

The use of cover crops in farmlands reduced the rate of evaporation of the irrigated water hence reducing the need for frequent irrigation of farmlands hence minimal energy applied.

The rotating of crops during the planting seasons reduces nutrient depletion in the soil and also improved the control of pests hence reduction of the energy used in the application of fertilizers and pest control herbicides (Basore et al., 2018).

The components in agriculture known to increase energy consumption is fertilizers, pesticides and other farm inputs. This is because they require energy for them to be manufactured and they also require energy for them to be applied. This form of energy consumption is referred to as indirect energy use in farms, more so the energy used in the manufacture of fertilizers, pesticides and the likes.

10.4 CONSERVATION OF AGRICULTURAL ENERGY

The stated techniques in the above section were are a summary of the practices which could help in the reduction of energy consumed in farmlands. The section below will elaborate more on ways that could be used to help in reducing the total amount of energy applied in agricultural activities. However, before we dive deep into the conservation techniques, we should first understand some of the commonly used termed in energy conservation.

The primary energy consumption in agriculture, (PEC), is known to be the total amount of energy consumed in any agricultural production system which includes basically the normal farm activities. It does also include the energy used in the production of all indirect farm inputs such as the likes of fertilizers, pesticides and insecticides.

The second term is the energy efficiency in agriculture improvement which is defined as the total reduction in the Primary Energy Consumption for only the energy used in the production of agricultural products and is expressed in weights per unit area, with the farm boundaries.

The third term is the Final Energy Consumption which refers to the total amount of energy supplied in the processing of the farm produce in the industries and factories which specialize in the processing of agricultural products. This type of energy excludes the energy required in the transportation of the farm produce from the farmland to either the factory or industry for the manufacturing process.

To conserve agricultural energy, several nations have come up with policies that ought to be followed by every farmer which also improves energy efficiency. Such policies help the nation in achieving its yearly targets of reducing energy consumptions, not only in the agricultural sector but also in other sectors which use different energy sources.

One of the policies is that of removing the obstacles in the energy market that lead to the availability of energy to farmers at high prices. The removal of such obstacles ensures that the supply of energy is efficient. The removal is done through the governments requesting energy companies to reduce their energy sales to a certain percentage annually. With the reduction in energy sale prices, farmers are further encouraged to manage their energy consumption since they have been done a favor.

The second policy is the establishment of structures that ensure minimal usage of energy. For instance, farmers are advised to construct buildings that are quite spacious for efficient air circulation and well insulated to provide the required temperature for storage of farm produce, without the use of any energy source.

Since most farms do make extensive use of heating facilities, most governments have come up with the policy that; heating companies be invented and assigned contracts to service several farms from a certain region with heating services hence the reduction in the total amount of energy consumed in farmlands. The last policy is that of frequent reporting of the amount of energy consumed during a certain period of time; analysis of whether the energy was conserved and in what ways was it conserved; analysis of the efficiency of energy used during a certain period among others. With such detailed reports, governments are able to track their progress in energy conservation and note which areas require the improvements.

The above policies improve energy efficiency and ensure that energy is well conserved. With conservation of energy, more so in the agricultural sector, the emission of greenhouse gases is reduced hence improving the quality of the product from farmlands, for the chances of them being contaminated by greenhouse gases are low.

10.5 ENERGY CONSERVATION IN AGRICULTURAL PUMPS

As stated earlier, the water pump system in most farmlands is the main consumer of agricultural energy. To conserve the energy used by most pumps; the pump size should be selectively selected depending on the land size to be irrigated. This ensures efficient use of energy sources depending on the power required to operate the pump.

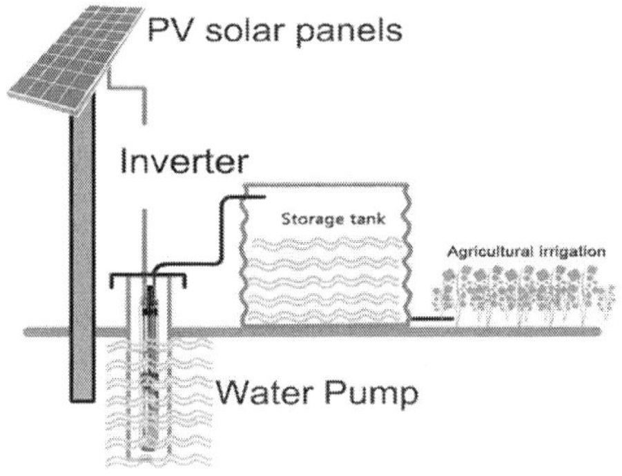

Figure 10.2: An image showing a solar-powered pump system.

Source: https://www.google.com/search?q=farm+pumping+system&client=ms-android-transsion-infinix-rev1&prmd=ivn&source=lnms&tbm=isch&sa=X&ved=2ahUKEwjHx73p7-nuAhWuz4UKHYfoAlQQ_AUoAXoECAMQAQ

The pump should also be bought depending on its source of water, whether it a well or a river or any other water reservoir. This is because the pump will use energy based on the area it draws water from. For instance, if it is a well, the pump might need lots of power to draw water and pump it to the farmland, which means it will use large amounts of energy.

The motor expected to power the pump should be selected based on the pump size. This is because selection of a motor size that does not fit with the pump size leads to the use of large amounts of energy during the powering of the pump.

The entire pump system should be well installed. From the connection of the pump to the motor to the connection of the pump to the water pipes; all this should be aligned well to ensure that the entire system functions properly hence the minimal amount of energy used to power the entire system. The water pipes should be connected in such a way that, they align with gravity for easier distribution of water throughout the farmland with minimal energy being used (Asif & Muneer, 2007).

The entire pump system should be periodically checked to ensure that all its components are working efficiently. For instance, the areas which experience frequent friction due to the movement of different parts should be frequently lubricated. The pipes which distribute water should be aligned appropriately to ensure that they distribute water evenly through the entire farmland. Warm out parts of the pumping system should be replaced with new ones. Generally, the entire pump system should be serviced regularly.

The pump system should be used efficiently by ensuring that the appropriate irrigation timetable is followed. This avoids over-irrigation of hence efficient use of water from the water source which avoids harming of crops and also wastage of energy sources. Crops that do not require large amounts of water should be irrigated using the drip method of irrigation which saves on water and energy used to supply the water to such points.

As some of the pumps are specific to certain energy sources, they ought to be powered only with that energy source. For instance, in water pumps that rely on diesel as their energy source, the engine output horsepower should be matched with a horsepower that supports the use of diesel fuel.

The distribution pipes should be kept at specific heights above the ground to ensure that they supply water efficiently. Bends should be used instead of elbows during the placement of the water distribution throughout the entire farmland. However, unnecessary bends should be avoided to allow for the smooth distribution of water.

Rain water harvesting is another technique which can be used in saving energy consumed in pumping system. Harvested water is mostly stored in structures such as tanks. All a farmer has to do when planning to irrigate the farmland; is open for the water to flow into the irrigation channels, which then distribute water throughout the entire farmland. Harvested rainwater does not have to be pumped hence saving on the energy consumed in powering the pump system.

10.6 ENERGY CONSERVATION IN AGRICULTURAL FIELD TRACTORS

Field tractors play a major role in making work easier for the farmers. For instance, fieldwork which required a lot of manual labor and a lot time for it to be completed; can be done within the shortest time possible and with minimal labor being applied, that is, with the help of a tractor.

However, for a tractor to complete its work, it requires energy to power its operations. The energy sources need to be used efficiently to avoid overconsumption that leads to pollution of the environment and an increase in the cost incurred to complete certain roles.

Figure 10.3: An image showing a farm tractor.

Source: https://www.google.com/search?q=field+tractors&client=ms-android-transsion-infinix-rev1&prmd=ivmn&source=lnms&tbm=isch&sa=X&ved=2ahUKEwj9g5H08OnuAhVkoXEKHSE5DhgQ_AUoAXoECAQQAg#imgrc=pU-m-cla_U82oM

To save on the energy used to power tractors, several measures have to be put in place. One of the measures is the purchase of the appropriate size of tractor depending on the work it is to do on the farmland. For instance, the tractor size required to carry out land clearing activities is not the same as the tractor required to do plowing. The appropriate tractor size saves on energy by ensuring that, the horsepower, that is the total amount of energy required to power the engine of a tractor, is minimal; hence saving on fuel. Statistics have shown that energy consumption increases with the use of excess tractor horsepower to carry out a job that requires minimal horsepower hence a reduction in energy efficiency. Still, the energy efficiency drops drastically if a small-sized tractor is used to carry out fieldwork that requires a large-sized tractor.

In some situations, all a farmer can afford is a tractor of the same size to carry out fieldwork which requires different horsepower. However, the good news is that the farmer can find ways to save on energy consumption, despite doing different tasks which require different horsepower. For instance, if the farmer is using a high horsepower tractor to carry out light farm activities; the farmer ought to run the tractor using a high gear but maintaining the engine speeds at a low level; which saves on fuel consumption hence energy conservation (Batterberry et al., 2010). The vice versa does also apply; that is, if the farmer has a low horsepower tractor and needs to complete heavy farm activities; he or she should run the tractor using a low gear at high engine speeds.

General maintenance of the tractor does also play a major role in saving up the energy consumed by a tractor. Farmers are advised to stay at the top of their game by always keeping the tractor clean, which is, frequently cleaning it up after farm activities; more so the air and fuel systems of a tractor ought to always be clean and free of any contaminants that might increase fuel consumption. Apart from cleaning, the entire tractor should be maintained regularly. The worn-out parts be replaced; parts that undergo lots of friction be frequently lubricated among other maintenance activities.

Most farmers, more so the large-scale farmers prefer that their tractors be refueled at home. So they set up farm tanks that will be refueled by fuel companies in large quantities, and then the farmers set up a refueling system whereby they can refuel their tractors whenever they need to be used. To increase the conservation of energy, farmers are advised to paint their home-based tanks with light colors, which instead of absorbing sunlight, reflect away sunlight; hence reducing the rates of evaporation, since fuels such as petrol and diesel are known to be volatile. Still, the home-based tanks are prone to robberies; so farmers are advised to label their tanks as red diesel which is known to be quite dangerous due to its high flammability rates. This discourages any potential thieves from attempting to steal the fuel from farmers saving on energy since it will be well used as it was budgeted.

All manufactured products, more so products manufactured for manual use do come with a manual guide. After the purchase of a tractor, farmers are encouraged to thoroughly study the tractor's manual guide and understand it; and then strictly adhere to all the rules and regulations on the use of the tractor. This avoids misuse of the tractor hence saving on the cost of energy consumed; a damaged tractor tends to consume more energy compared to a tractor that is well maintained.

During harvesting, the heavy load tractors should not be used. This is because the addition of the farm produce increases the total weight of the tractor hence an increase in the rates of fuel consumption. Farmers are advised to use lightweight vehicles, to transport their harvest from the farm to the storage unit or to the factory. This should also apply to the harvesting tractors. The harvest- consuming tractors are not supposed to be the ones transporting the produce they harvest from the farmland to the storage units, rather, they should only harvest and leave the produce to the lightweight tractors or vehicles to transport it to the storage unit. This saves on cost because, lightweight tractors consume little energy (Badawy et al., 2010).

If possible, farmers should minimize on the movement of tractors on the roads. Field tractors should only be used during field operations, and if possible, not used for any other purposes. Such maintenance saves on frequent wear and tear of the tractor's spare parts; and with minimal wear and tear, the tractor will consume less energy for all its components are functioning efficiently. In addition to this, tractors should be serviced from time to time to ensure that they are in perfect condition for farm operations.

10.7 ENERGY CONSERVATION IN FARM IRRIGATION SYSTEM

The irrigation system is another important system of almost all agricultural fields, and it is independent of the pumping system. As stated earlier, the pumping system is mainly responsible for drawing water from the water source, and channeling it to the water channels responsible for Irrigation; whereas, the irrigation system is solely responsible for distributing water from the pump system throughout the entire farmland.

Figure 10.4: An image showing an example of irrigation technology, the Sprinkler irrigation.

Source: https://www.google.com/search?q=farm+irrigation+system&client=ms-android-transsion-infinix-rev1&prmd=ivn&source=lnms&tbm=isch&sa=X&ved=2ahUKEwigtuGD8unuAhVBt3EKHeHxDc0Q_AUoAXoECAQQAQ

For an irrigation system to function properly, it does require energy to power its operations. This energy ought to be conserved to save on the general costs of maintaining farmland.

One of the ways in which energy consumed in Irrigation systems can be conserved is through water management. Water management can be done by lining a specific number of watercourses which could also be known as irrigation channels. With irrigation channels, the farmer is able to ensure that water is distributed to all key components of the farm. With this, pumping will be done according to the timetable, and irrigation will also be done according to the timetable, hence eliminating the need for unnecessary irrigation that in turn saves up on the energy used for both irrigation purposes and pumping water.

To avoid excessive use of water during irrigation, farmers could use the latest technologies in irrigation which ensure that, just the required amount of water is used up and that energy is saved (Cardenas et al., 2016).

One of the irrigation technologies is drip irrigation that is also known as trickle irrigation. With drip irrigation, water pipes with thing holes which are water passages are lined up in the entire farmland. Once the water is pumped and spread through the irrigation channels, the drip pipes allow water to get to the plant drop by drop. The water drops can either be onto the soil surface or deep into the plant roots. The water makes its way to the routes through and emitters which are connected to the drip system to supply water deep into the plant roots drop by drop. The good thing about drip irrigation is that it saves on water. Drip irrigation does also preserves the soil by providing minimal to almost no disturbance to the soil particles. It does also have the advantage of not affecting the working of fertilizers on plants by ensuring that, fertilizers spread on the topsoil is not scattered all over; and that it remains intact on its point of application working effectively as intended.

The second irrigation technology is sprinkler irrigation. Sprinkler irrigation works using the mechanisms of rainfall. Water from the pumping system is passed onto the irrigation system which supplies it through the entire farm using pipes. The unique thing about sprinkler irrigation is that; water is distributed to the crops through sprinklers. The sprinklers are connected to the pipes and raised at a higher ground such that their role is spraying water onto the crops. The speeds of water can be controlled. For sprinkler irrigation to function well, both the pumping system and the irrigation system should be working properly to ensure that water gets to the sprinkler for it to be used to irrigate the crops. The advantage with sprinkler irrigation is that, not only the roots of the plant will be nourished, but also its leaves for some water will fall on them creating that cool temperature around them hence their nourishment.

The third irrigation technology is referred to as the Hydroflume irrigation system. As the name suggests, Hydroflume irrigation incorporates the use of a flume. A flume is basically a man-made water channel. It can be made from either wood or cement or metal. However, for durability purposes and to avoid contamination of the water used for irrigation by having rust on the flume, plastic flumes have been made. The plastic flumes are made from strong plastic materials. The water flume diverts water from a certain water reservoir directing it to the farmland. Still, water could be pumped into the flume for it to distribute the water to the farmland (Ceylan et al., 2004). The flume is made in a way that it has an opening that helps in distributing water throughout the farm. The openings are such that, they are placed fully adjacent to open ditches manually made throughout the farm. Once the water has been pumped, and the flume supplied with enough water via the irrigation hose pipe, the openings open up and feed the open ditches with water. The open ditches then distribute water through the entire farmland. With Hydroflume irrigation, water distribution is controlled hence the usage of water at low rates which saves up on energy used to supply the Irrigation hose pipe with water.

10.8 ENERGY CONSERVATION THROUGH PRESERVATION OF THE SOIL

As stated earlier, a lot of energy is used up in the manufacture of farm inputs such as the likes of fertilizers. Still, the process of applying fertilizers to our farmlands is also known to consume a lot of energy. For energy to be conserved, the soil should be preserved such that it will be fertile and with no need for the application of fertilizer to improve its fertility rates.

One of the ways in which the soil can be preserved is by the use of the Zero Tillage Mechanisms. This mechanism ensures that the soil is not disturbed in any way, and if there is any disturbance, it is disturbed as little as possible. This is made possible by directly planting the crop directly into the seedbed; instead of having to plant it on the nursery and later transplant it onto the main farmland. The seedbed should not have been tilled since the harvesting process of the previous crop planted on it.

Apart from the advantage of retaining the soil fertility; with zero tillage mechanism, the irrigation rates are reduced. This is because the soil will have retained its moisture for it to experience little to no disturbance. Instead of regularly irrigating, this moisture plays a big role in helping the directly planted seed to germinate. The moisture does also nature the plant to a certain level.

Soils should not be overly irrigated to ensure that they maintained within the same locality. With over-irrigation, chances of soil erosion are high which in turn lowers the soil fertility by causing disturbance to the soil.

Farmers who practice mixed farming, that is both the rearing of livestock and the growing of crops, are encouraged to use organic manure instead of fertilizers, in order to boost soil fertility. Organic manure is known to last in the soil for quite a long time, so the farmers will not have to keep on applying manure over and over again, gene saving on fertilizer application energy.

One of the main advantages of no-tillage mechanism is that it saves the farmer a bigger percentage of the crop production cost. For instance, energy will be conserved by not fueling the tractors to till the land. Generally, the no-till mechanisms require minimal labor, do not consume a lot of the farmer's time and no energy form is used to power its operations (Cook, 2021).

The other technique of conserving energy Consumption in soils is by preserving the topsoil, which is the first layer of the soil. Topsoil can be preserved through the spreading of soil cover, commonly known as mulch. The application of mulch on the topsoil reduces the rates of evaporation of the groundwater for the mulch will provide more of a shade to the topsoil. As time goes by, the mulch does get to the point of decay. The decaying of mulch produces high organic content matter which boosts the soil fertility saving energy consumption for it eliminates the need for fertilizer application. The topsoil is known to harbor some tiny micro-organisms. Such microorganisms are well protected and provided with a serene environment for their growth and reproduction process if the soil is covered in the mulch. This is because, mulch boosts the soil temperatures, providing the microorganisms with just the perfect temperatures for survival. Soil microorganisms are known to play a bigger role in aerating the soil. Well-aerated soil saves energy consumption in farms for it reduces the need of having to frequently till the land; which is a process that requires energy for it to be completed. Mulch does also play an important role in ensuring that the topsoil is covered hence discouraging the occurrence of any form of soil erosion. Mulch is also known to be quite effective in weed control saving on energy used during the weeding process. This is because mulch is known to suppress the weeds hence discouraging their growth.

To save on the energy used in the manufacture of farm inputs such as the likes of pesticides, and the energy used in the application of pesticides; farmers are advised to practice crop rotations. With crop rotations, pests and organisms causing crop diseases are discouraged to move in circles during the different seasons of the plantation. Such organisms tend to hide in the soil after the harvesting season. Once the crops are rotated from one planting season to another; a pest or disease-causing microorganisms will have to die for most of them are crop-specific, and the new crop does not give room for them to grow and attack it. The same does apply to weeds, with crop rotations, most weeds will be suppressed by the new crop variety hence saving on energy used in the manufacture herbicides and also the energy used to apply the herbicides.

The land sparing technique, however difficult it might seem to be, more so in our current revolutionary age since the demand for land has increased tremendously; does for sure save on energy consumed in improving soil fertility. Land sparing allows a farmer to plow on a piece of land in one planting season; and in the second planting season, the farmer moves to a different piece of land leaving the original piece of land bare. The bare land gets the opportunity to replenish its mineral contents hence boosting its fertility. Come the next planting season, after a piece of land, had been left only for the growth of natural vegetation; will be quite fertile eliminating the need for application of fertilizer or manure, which is a practice that saves on energy consumed in practices aimed at boosting soil fertility.

10.9 CONSERVATION OF AGRICULTURE ENERGY BY USE OF GREEN FUEL

Most of the farm activities are powered using fuels, and most of the fuels which have been in use in agricultural fields are volatile. By volatile it means that they get deported easily for large volumes of volatile fuels are required to carry out a small task. However, the good news is that green fuels have been discovered and it's slowly becoming popular in most agricultural fields (Cullen et al., 2011).

One of the examples of green fuels is the biodiesel. Biodiesel is a type of fuel harnessed from vegetables or animals. Biodiesel is made up of several chemical elements that include methyl, propyl and ethyl, all of which are of the chemical group of the esters. The production process of biodiesel fuels involves the reaction of fats oils; that is oil obtained from vegetables and fats obtained from animals. To the mixture of fats and oils, some small amount of alcohol is added for biodiesel fuels to be fully made. The main advantage of biodiesel is that it does not pollute the environment and it is not that volatile; meaning that it can be used in small amounts to carry out a heavy workload.

The other green fuel is referred to as bioethanol. As the name suggests, bioethanol fuels are made from organic materials that are later mixed with elements containing some percentage of ethanol. Basically, bioethanol fuel is made from agricultural crops such as sugarcane and biomass. Bioethanol fuels are said to be free of any carbon dioxide in them; hence they are advantages for they do not pollute the environment by adding more carbon dioxide to the atmosphere. The fact that bioethanol fuels are free of carbon dioxide means that they are not volatile hence they can be used in small amounts to complete a heavy workload.

Green fuel can also be obtained through the use of wind energy. Wind energy eliminates the need of having to use fossil fuels to power Agricultural activities. Wind energy can be harnessed through the installation of wind turbines. Wind turbines are known to convert the energy from the wind to electricity through the rotation of propeller-like structures fitted on top of the wind turbine. The propeller blandest are turned by the wind, and their speed depends on the speed of the wind. With an increase in the speed of the rotation of the propellers, a structure on the wind turbine which is referred to as a rotor is turned so as to generate mechanical energy. The mechanical energy is then used to turn the drive shaft which is powers in the electric generator hence the production of electricity. The main advantage of wind energy is that it is cheap and improves on energy consumed since wind energy is a natural and free source of energy that is also renewable. The only cost that comes with energy is that of installation of wind turbines, purchase of electric generator and the cost incurred during the maintenance of the generator and the wind turbine (Chowdhury et al., 2020). Furthermore, the drive shaft from the wind turbine can be directly connected to the pumping water system motor, hence the generation of energy required to pump water to the irrigation channels. These sales on the high energy consumption rates of fossil fuels such as the likes of diesel are used.

Figure 10.5: An image showing a wind plant with several wind turbines.

Source: https://www.google.com/search?q=wind+energy&client=ms-android-transsion-infinix-rev1&prmd=inv&source=lnms&tbm=isch&sa=X&ved=2ahUKEwjdhY_X9OnuAhVyyoUKHRa-BAAQ_AUoAXoECCEQAg

The other form of green fuel is solar energy. Solar energy is known to be the light and heat radiated from the sun onto the earth's surface. This heat is harnessed through various technologies which have been invented so as to convert to electricity to be used for various farm activities. The conversion is done either directly using the technology of photovoltaics (PV) or indirectly using the technology of concentrated solar power (CSP). The photovoltaic technology is known to convert light from the sun into electricity using the photoelectric effect whereas the concentrated solar power, as the name suggests; is known to employ the use of lenses that focus sunlight from a large atmospheric area into a small beam so as to generate electricity. Solar energy can be used and has been used in conserving the energy consumed in our farms. For instance, solar energy can be used in the drying of crops saving on the energy consumed in powering grain drying equipment. Electricity produced through solar energy can be used to power the pump system. Before manure is spread onto the farm, farmers are advised to dry it up; and solar energy plays a big role in drying up manure before it is spread on the farm. In greenhouses, solar energy can be converted into heat so as to improve the yearly production of crops grown in greenhouses by boosting their growth rate. Solar energy can be used in the manufacture of fuels that are more durable and effective as compared to fossil fuels (De Andrade et al., 2015). This is possible through the use of solar energy in driving solar chemical processes which can either be thermochemical (using the sun's heat) or photochemical (using the light from the sun). Such fuels can be used as a source of energy for several farm activities hence saving on total energy consumption on the farm.

Figure 10.6: An image showing the process of harnessing solar energy.

Source: https://www.google.com/search?q=solar+energy&tbm=isch&ved=2ahUKEwjq446E8-nuAhXEgHMKHYKEAMkQ2-cCegQIABAC&oq=solar+energy&gs_lcp=ChJtb2JpbGUtZ3dzLXdpei1pbWcQAzoCCAA6BggAEAcQHjoGCAAQCBAeUPiuB1izwAdg38cHaAFwAHgAgAG6FogBuhaSAQM5LTG-YAQCgAQGwAQDAAQE&sclient=mobile-gws-wiz-img&ei=ml0pYOquEsSBzgOCiYLIDA&bih=592&biw=360&client=ms-android-transsion-infinix-rev1&prmd=inv

Biogas is another form of green fuel that can be used in agricultural fields. It plays a big role in saving the total amount of energy consumed. Furthermore, it is made from raw materials which are easily obtained within the farm. For methane gas to be produced, a biogas plant ought to be set up. The plant is mainly used to digest methane from both plant and animal waste. Once the methane gas has been harvested, it is referred to as the Biogas. Biogas is distributed through the farm to key points which require energy. For instance, biogas is transferred to the farm's kitchen and then used to power the cooking system and help in cooking activities; which saves on energy used in preparing meals for farmers in the farmland. Biogas can also be used to power the lighting system in a farm saving on the energy used for lighting purposes.

Conservation of Agricultural Energy through Installation of Wind-breakers.

Most individuals look down on the importance of growing Windbreakers on their farms. Research has shown that agricultural fields that have grown several windbreakers within their surroundings have a high percentage of energy efficiency.

Figure 10.7: An image showing a farm field with windbreakers.

Source: https://www.google.com/search?q=farm+windbreakers&tbm=isch&ved=2ahUKEwii_77U9enuAhXF04UKHaTXDAQQ2-cCegQIABAC&oq=farm+windbreakers&gs_lcp=ChJtb2JpbGUtZ3dzLXdpei1pbWcQA1C4S1iYVW-D4WmgAcAB4AIABAIgBAJIBAJgBAKABAcABAQ&sclient=mobile-gws-wiz-img&ei=W2ApYOLlLMWnlwSkr7Mg&client=ms-android-transsion-infinix-rev1&prmd=inv

One, windbreakers help in the prevention of soil erosion. The chances of soil being eroded, more so through the wind agent of erosion, are low with the availability of Windbreakers. This is because the windbreakers reduce the speed of the wind to a level where the wind has no power to lift up soil particles from the ground and transport them to another region. This saves on the energy used to improve soil fertility which had been lowered due to soil erosion.

Windbreakers are also known to reduce the rates of evapotranspiration on the crops. This is because the windbreakers provide shade to the crops sheltering from sunlight; hence prevention of evaporation of groundwater into the atmosphere and also prevention of water loss from the crops leaves and any other stomata section. This saves on the energy used during frequent pumping of water for irrigation purposes.

Plants are said to always be in constant need of humidity and calm air. With the presence of Windbreakers, the speeds of wind around the farm surroundings are lowered; hence providing high humidity conditions within the farm and calmness in the farm. High humidity and calm air allow the crops to use water more efficiently; hence saving on the energy used in pumping water to irrigate the crops. With windbreakers, crops that are susceptible to wind destruction are preserved hence saving on the energy used for the replanting of the already damaged farm crops.

Most farms are located adjacent to each other. And the farming practices practiced on adjacent farms maybe be different. One of the farms may specialize in purely organic farming practices whereas the second farm specializes in non-organic farming. With windbreakers, the mixing of the organic farm inputs with non-organic farm inputs is prevented for the windbreakers weaken the wind hence it is not able to transport the farm inputs from one farm to another. This saves on the energy used for the rectification of the effects caused by the mixing of farm produce from adjacent farms.

Still, when it comes to windbreakers, they tend to provide a safe environment for the growth of crops. Most crop pollinators are said to be attracted to such safe environments hence increasing the chances of pollination which increases crop yield. This saves on energy consumed for a farmer is guaranteed of reciprocated produced for all the energy that was invested in the production of the crop.

The incorporation of windbreakers is of utmost importance in increases the rates of energy conservation in agriculture. Farmers are advised to at least plant a few trees around their farms and in between so as to act as windbreakers. This will reduce the negative effects caused by wind and also save on energy consumption in farms.

Challenges facing Conservation of Agricultural Energy

Conservation of Agricultural Energy is faced by a few challenges; some of which seem to be so easy to solve, whereas others might take some time before they can be completely solved (Delmastro & Gargiulo, 2020).

One of the challenges facing energy conservation in agriculture is the high initial cost that is required to aid in the setting structures and equipment that help in energy conservation. For instance, the setting up of a biogas plant requires a lot of capital which might not be available to most farmers. The setting up of a solar energy harnessing plant is also expensive. For a farmer to purchase a wind turbine, it requires a lot of capital which might not even be readily available and, of course the farmer will not only need one turbine, several for the required energy to be generated. Still, on wind energy, the farmer might need to have a generator that will help in the conversion of the mechanical energy from the rotor to electrical energy, which is an added cost to them.

For a farmer to incorporate agricultural activities aimed at conserving energy consumption, he or she ought to be skilled. The acquisition of such skills is time-consuming, and it requires capital for the farmers to be educated on the agricultural methods aimed at conserving energy consumption.

The other common problem, which is easy to solve, yet has become a bigger problem is the fact that most farmers are so much used to conventional farming practices, which lead to high energy consumption. Farmers are so reluctant to move to conservation agriculture which seems so hard to them yet it is so simple once one has fully adapted to it.

With the effort of trying to incorporate practices that promote energy conservation; farmers tend to face loss at the beginning of the process. This might be a discouragement to most of them. Hence, the availability of people who are financially stable to bear the first few losses after shifting to conservation agriculture is low.

Most of the practices which promote energy conservation, seem to be an irony in a way. This is because, with the cutting of the use of fertilizers, the use of herbicides for weeding, the use of tractors for tilling by opting for the no-tillage mechanisms; the rate of crop production drops. This brings the question of; should nations starve themselves by trying to conserve energy or should ignore energy observation and feed.

The cost of maintenance of most farm systems is high. For instance, the frequent servicing of tractors, pumping systems, irrigation channels among others costs the farmer. The repairing of broken parts, replacement of wrong parts of such system, does cost the farmer which in most cases acts as a discouragement to farmers; pushing most of them to be reluctant in maintenance activities and repairing activities.

Generally, the incorporation of energy-saving techniques in the field of agriculture seems to be facing a lot of challenges; which make it seem impossible yet in a way it is also possible.

CONCLUSION

The conservation of agricultural energy is so important. This is because most nations rely on Agriculture for the boosting of their economy and also for the survival of people. The challenges facing energy Conservation ought to be addressed. For instance, the issue to do with ignorance among farmers should be addressed by the preparation of several workshops to educate farmers on the importance of conserving energy, not only to agriculture but to the environment in general.

Most of the equipment that helps in energy conservation should be lowered their market price. If possible, farm Sacco should be formed to help farmers in bulky purchase of the equipment they need most which in a way reduces the cost and makes it easier for farmers to access such equipment. As for structures such as solar plants or wind plants or even biogas plants; farm within the same locality should be encouraged to come together and set up a common plant where they would draw their energy hence saving on the cost of having to set up an independent plant in each farmland (Edelenbosch et al., 2020).

The government should support farmers who wish to use farming practices which save on energy. This increases the number of individuals practicing conservation agriculture, which not only save energy consumption in the Agricultural sector; but also in the entire nation.

As for the issue of windbreakers, farmers should be educated on the ways in which windbreakers are to be incorporated into their farmlands to save on energy consumption.

Let us all have a positive approach to conservation agriculture, and the benefits reaped will be many.

References

1. Armaroli, N., &Balzani, V. (2007). The future of energy supply: challenges and opportunities. *AngewandteChemie International Edition*, *46*(1-2), 52-66.
2. Anishin, A. S., Piriashvili, B. Z., &Chirkin, B. P. (2013). Trends and prospects of ensuring energy resources in Ukraine. *Studies on Russian Economic Development*, *24*(1), 84-98.
3. Ahmad, T., & Zhang, D. (2020). A critical review of comparative global historical energy consumption and future demand: The story told so far. *Energy Reports*, *6*, 1973-1991.
4. Basore, P. A., & Cole, W. J. (2018). Comparing supply and demand models for future photovoltaic power generation in the USA. *Progress in Photovoltaics: Research and Applications*, *26*(6), 414-418.
5. Asif, M., & Muneer, T. (2007). Energy supply, its demand and security issues for developed and emerging economies. *Renewable and sustainable energy reviews*, *11*(7), 1388-1413.
6. Batterberry, T., Miller, M., Jaskolka, K., & Toll, R. (2010). *U.S. Patent Application No. 12/483,975.*
7. Badawy, R., Hirsch, B., & Albayrak, S. (2010). Agent-based coordination techniques for matching supply and demand in energy networks. *Integrated Computer-Aided Engineering*, *17*(4), 373-382.
8. Cardenas, L. M., Franco, C. J., &Dyner, I. (2016). Assessing emissions–mitigation energy policy under integrated supply and demand analysis: the Colombian case. *Journal of Cleaner Production*, *112*, 3759-3773.

9. Ceylan, H., & Ozturk, H. K. (2004). Estimating energy demand of Turkey based on economic indicators using genetic algorithm approach. *Energy Conversion and Management*, *45*(15-16), 2525-2537.
10. Cook, M. (2021). Trends in global energy supply and demand. In *Developments in Petroleum Science* (Vol. 71, pp. 15-42). Elsevier.
11. Cullen, J. M., Allwood, J. M., &Borgstein, E. H. (2011). Reducing energy demand: what are the practical limits?. *Environmental science & technology*, *45*(4), 1711-1718.
12. Chowdhury, M. S., Rahman, K. S., Selvanathan, V., Nuthammachot, N., Suklueng, M., Mostafaeipour, A.,&Techato, K. (2020). Current trends and prospects of tidal energy technology. *Environment, Development and Sustainability*, 1-16.
13. De Andrade, J. B. S. O., Dutra, L., Schwinden, N. B. C., & De Andrade, S. F. (2015). Future scenarios and trends in energy generation in Brazil: supply and demand and mitigation forecasts. *Journal of Cleaner Production*, *103*, 197-210.
14. Delmastro, C., & Gargiulo, M. (2020). Capturing the long-term interdependencies between building thermal energy supply and demand in urban planning strategies. *Applied Energy*, *268*, 114774.
15. Edelenbosch, O. Y., Van Vuuren, D. P., Blok, K., Calvin, K., & Fujimori, S. (2020). Mitigating energy demand sector emissions: The integrated modelling perspective. *Applied Energy*, *261*, 114347.
16. Elghawi, U., & El-Osta, W. (2015). The alternative energy sources and technologies suitable for Libyan future energy demand mix. *International Journal of Energy Technology and Policy*, *11*(1), 36-52.
17. ErselCanyurt, O., Ceylan, H., Kemal Ozturk, H., &Hepbasli, A. (2004). Energy demand estimation based on two-different genetic algorithm approaches. *Energy Sources*, *26*(14), 1313-1320.
18. Esen, Ö. (2016). Security of the energy supply in Turkey: Prospects, challenges and opportunities. *International Journal of Energy Economics and Policy*, *6*(2).
19. Ghalehkhondabi, I., Ardjmand, E., Weckman, G. R., & Young, W. A. (2017). An overview of energy demand forecasting methods published in 2005–2015. *Energy Systems*, *8*(2), 411-447.

20. Grunewald, P., &Diakonova, M. (2018). Flexibility, dynamism and diversity in energy supply and demand: A critical review. *Energy Research & Social Science*, *38*, 58-66.

21. Goiri, Í., Haque, M. E., Le, K., Beauchea, R., Nguyen, T. D., Guitart, J., ... &Bianchini, R. (2015). Matching renewable energy supply and demand in green datacenters. *Ad Hoc Networks*, *25*, 520-534.

22. Hamlet, A. F., Lee, S. Y., Mickelson, K. E., & Elsner, M. M. (2010). Effects of projected climate change on energy supply and demand in the Pacific Northwest and Washington State. *Climatic Change*, *102*(1), 103-128.

23. Hossain, M. F., Hossain, S., & Uddin, M. J. (2017). Renewable energy: Prospects and trends in Bangladesh. *Renewable and Sustainable Energy Reviews*, *70*, 44-49.

24. Huang, Y., Bor, Y. J., & Peng, C. Y. (2011). The long-term forecast of Taiwan's energy supply and demand: LEAP model application. *Energy policy*, *39*(11), 6790-6803.

25. Liu, Z., Yu, H., & Liu, R. (2019). A novel energy supply and demand matching model in park integrated energy system. *Energy*, *176*, 1007-1019.

26. Liu, J., Niu, D., & Song, X. (2013). The energy supply and demand pattern of China: A review of evolution and sustainable development. *Renewable and Sustainable Energy Reviews*, *25*, 220-228.

27. Liuyijuna, L. (2011). Analysis of China's current energy structure and trend of development: supply and demand. *Energy Procedia*, *5*, 2593-2598.

28. Majidi, C. (2014). Soft robotics: a perspective—current trends and prospects for the future. *Soft robotics*, *1*(1), 5-11.

29. Mortazavi, S. M., Maleki, A., &Yousefi, H. (2019). Analysis of robustness of the Chinese economy and energy supply/demand fluctuations. *International Journal of Low-Carbon Technologies*, *14*(2), 147-159.

30. Moran, C. J., & Kunz, N. C. (2014). Sustainability as it pertains to minerals and energy supply and demand: a new interpretative perspective for assessing progress. *Journal of cleaner production*, *84*, 16-26.

31. Muralitharan, K., Sakthivel, R., &Vishnuvarthan, R. (2018). Neural network based optimization approach for energy demand prediction in smart grid. *Neurocomputing*, *273*, 199-208.
32. Mukhopadhyay, S., &Nateghi, R. (2017, July). Estimating climate—demand nexus to support longterm adequacy planning in the energy sector. In *2017 IEEE Power & Energy Society General Meeting* (pp. 1-5). IEEE.
33. Ouedraogo, B. I., Levermore, G. J., & Parkinson, J. B. (2012). Future energy demand for public buildings in the context of climate change for Burkina Faso. *Building and Environment*, *49*, 270-282.
34. Oughton, E., Frias, Z., Russell, T., Sicker, D., &Cleevely, D. D. (2018). Towards 5G: Scenario-based assessment of the future supply and demand for mobile telecommunications infrastructure. *Technological Forecasting and Social Change*, *133*, 141-155.
35. Peacock, A. D., Owens, E. H., Roaf, S., Corne, D. W., Dissanayake, M., Tuohy, P. G., ... & Galloway, S. (2014, December). Autarkic energy systems: balancing supply and demand with energy storage and controls in local energy micro-grids. In *2014 Asia-Pacific Solar Research Conference.*
36. Perwez, U., Sohail, A., Hassan, S. F., & Zia, U. (2015). The long-term forecast of Pakistan's electricity supply and demand: An application of long range energy alternatives planning. *Energy*, *93*, 2423-2435.
37. Rosenberg, E., Lind, A., &Espegren, K. A. (2013). The impact of future energy demand on renewable energy production–Case of Norway. *Energy*, *61*, 419-431.
38. Raymond, L., Gotham, D., McClain, W., Mukherjee, S., Nateghi, R., Preckel, P. V., &Wachs, E. (2019). Projected climate change impacts on indiana's energy demand and supply. *Climatic change*, 1-15.
39. Schaeffer, G. J. (2015). Energy sector in transformation, trends and prospects. *Procedia Computer Science*, *52*, 866-875.
40. Simoes, S., Seixas, J., Fortes, P., &Huppes, G. (2014). The savings of energy saving: interactions between energy supply and demand-side options—quantification for Portugal. *Energy Efficiency*, *7*(2), 179-201.
41. Shove, E., & Walker, G. (2014). What is energy for? Social practice and energy demand. *Theory, Culture & Society*, *31*(5), 41-58.
42. Sørensen, B. (2008). A sustainable energy future: Construction of demand and renewable energy supply scenarios. *International Journal of Energy Research*, *32*(5), 436-470.

43. Silvente, J., Kopanos, G. M., Pistikopoulos, E. N., &Espuña, A. (2015). A rolling horizon optimization framework for the simultaneous energy supply and demand planning in microgrids. *Applied Energy*, *155*, 485-501.
44. Silvente, J., Graells, M., Espuña, A., & Salas, P. (2012, September). An optimization model for the management of energy supply and demand in smart grids. In *2012 IEEE International Energy Conference and Exhibition (ENERGYCON)* (pp. 368-373). IEEE.
45. Utama, N. A., Fathoni, A. M., Kristianto, M. A., & McLellan, B. C. (2014). The end of fossil fuel era: supply-demand measures through energy efficiency. *Procedia Environmental Sciences*, *20*, 40-45.
46. Uchiyama, Y. (2002). Present efforts of saving energy and future energy demand/supply in Japan. *Energy Conversion and Management*, *43*(9-12), 1123-1131.
47. van Ruijven, B. J., De Cian, E., & Wing, I. S. (2019). Amplification of future energy demand growth due to climate change. *Nature communications*, *10*(1), 1-12.
48. Wang, Y., Gu, A., & Zhang, A. (2011). Recent development of energy supply and demand in China, and energy sector prospects through 2030. *Energy Policy*, *39*(11), 6745-6759.
49. Waibel, C., Evins, R., &Carmeliet, J. (2019). Co-simulation and optimization of building geometry and multi-energy systems: Interdependencies in energy supply, energy demand and solar potentials. *Applied Energy*, *242*, 1661-1682.
50. Weiss, M., Junginger, M., Patel, M. K., & Blok, K. (2010). A review of experience curve analyses for energy demand technologies. *Technological forecasting and social change*, *77*(3), 411-428.
51. York, R. (2016). Decarbonizing the energy supply may increase energy demand. *Sociology of Development*, *2*(3), 265-272.
52. Yoon, H. (2018). A review on water-energy nexus and directions for future studies: From supply to demand end. *Documents d'anàlisigeogràfica*, *64*(2), 365-395.
53. Züttel, A., Remhof, A., Borgschulte, A., & Friedrichs, O. (2010). Hydrogen: the future energy carrier. *Philosophical Transactions of the Royal Society A: Mathematical, Physical and Engineering Sciences*, *368*(1923), 3329-3342.

INDEX